# RUBBER SEED:

# AN AFRICAN SCIENTIST IN THE AMAZON BASIN

**Kome Onokpise**

**SokheChapke Publishing**

Published by:
SokheChapke Publishing, Inc., ©
P.O Box 21161
Tallahassee, FL 32316
545 East Tennessee St., Suite 200-A
Tallahassee, FL 32308
Telephone: 866-711-5984
Email: info@sokhechapkepublishing.com
Website: sokhechapkepublishing.com.



ISBN: 978-0-9833991-0-0
Published and printed in the United States of America

*Library of Congress Card Catalog Data*

# TABLE OF CONTENTS

## *DEDICATION*

This book is dedicated to the following:

1. My late father, Mr. Jerome Esagwu Onokpise Anyisire and my late mother, Mrs. Margaret Evrahwoma Onokpise Anyisire.
2. My team mates in the Amazon Basin, Brazil for "Germplasm 81", and to all the scientists who have and continue to conduct research on one of the most important renewable natural resource with which humans have been blessed.
3. To the many rubber tappers who toil in the rubber tree plantations and in the wild to obtain the latex which drives and flies the world.

# ACKNOWLEDGEMENTS

I wish to thank God Almighty who makes all things possible and without whom I may not have been selected let alone participate in an international expedition worth a lifetime experience. The final title for this book, "Rubber Seed -----" came about during a discussion with my brother, Mr. Patrick Onokpise, while I spent four days with him, his wife and children as part of an official visit to Nigeria from South Africa enroute to Florida. I gratefully acknowledge him for his insight in suggesting a memorable title for the book. Special acknowledgement go to my immediate and extended families starting with my beloved wife, Mrs. Lucy Omotaka Onokpise, who with a pregnancy of five months was very patient to see the planning process for the expedition and after giving birth four months later was very happy to see me undertake this expedition. Thanks to my late mother, Mrs. Margaret Onokpise-Anyisire, for all of her usual parental and grandparental support who at the time had to shuttle between Benin City and Lagos, Nigeria to assist my sister and I in looking after her grand children, while always encouraging me to be the best son in the world. I thank my sister, Mrs. Josephine Akiti, who ensured that I had a pleasant departure for and arrival from Brazil amidst all the events that took place throughout those months including my ability to wade through the painful traffic of Lagos. Thanks are also extended to her husband, Mr. Ernest Akiti for all his support. My special appreciation goes to my children, Oghenemaro, Omurhu, Oghogho and Onoriode who have continuously made dad work and reminded him of his fatherly responsibilities. It is for our children and grandchildren for whom endeavors like the one written here even become more rewarding.

The Rubber Research Institute of Nigeria (RRIN) and the Nigerian Federal Ministry of Science and Technology provided funding for my travel out of and return to Nigeria. This financial support is gratefully acknowledged. Special thanks are extended to the International Rubber Research and Development Board (IRRDB) and the Brazilian Ministry of Science and Technology, for funding the entire Germplasm 81 *Hevea brasiliensis* expedition. I am particularly grateful for the on site support throughout the expedition which was provided and coordinated by these important Brazilian agencies: the National Center for Genetic Resources (CENARGEN), the Brazilian Enterprise for Agricultural Research (EMBRAPA) and the Brazilian Superintendence for Rubber Development (SUDHEVEA). The excellent organization and coordination ensured that we as scientists had a very safe expedition especially in some extremely remote parts of the Amazon Basin. I like to extend special acknowledgements to all my team mates in Mato Grosso especially Dr. Paiva, our team leader, for their wonderful friendship, dedication to duty and selfless assistance and patience even during those challenging moments in the remotest parts of the Amazon jungle. Their professionalism at every level made for the tremendous success of our team.

I am particularly grateful to Dr. Brian McBlain and his wife, Kathy McBlain, and Dr. Gary Hokstra and his wife, Cindy Hokstera, my Canadian family friends who as course mates created an atmosphere that gave me the confidence to return to Canada as part of the expedition in terms of quarantining in a temperate country. Once in Canada, I met a few more friends that made my one-week stay in Canada, very rewarding and memorable. My thanks also go to the bank manager at the Canadian Imperial Bank of Commerce for tremendous assistance with the purchase of my car. The entire quarantine period was indeed not only relaxing but also very fulfilling. Thus while fulfilling the IRRDB scientific requirements, I was able to meet some other personal goals. Hence I deeply acknowledge the support of all my Canadian friends and acquaintances.

I like to thank Dr. Cyril Ayeke and Dr. E. K. Okaisabor, former Directors of the RRIN. Dr. Ayeke sent me to attend the first "Workshop on International Collaboration in Hevea

Breeding and the Collection and Establishment of Materials from the Neo Tropics." Attendance at this workshop gave me a first hand knowledge and enabled me to have an insight to the preparations for the expedition in Brazil. Dr. Okaisabor was very kind to select me for the actual germplasm collection in Brazil. It was a bold action considering I was not only the youngest of three scientists in the Plant Breeding Division (PBD) of RRIN at the time, I had just returned from Canada after obtaining my M.Sc. degree in Plant Breeding and Genetics. Special thanks are extended to Dr. H. Mekako who as head of the PBD, RRIN worked very closely with me to ensure that my research projects were fully maintained and the preparations for the expedition at the institute's level were effectively carried out as instructed by Dr. Okaisabor. The efforts of my field technicians, Mr. Emmanuel Aransomwan and Mr. Clifford Annetor for their dedication in establishing and maintaining the seedling nursery for the rubber seeds (*Hevea benthamiana* and *Hevea spruceana*) other than *Hevea. brasiliensis* which I brought from the Amazon Basin. My thanks are extended to my good friend at RRIN, Mr. Ike Ugwa for all his support for my family during my four months away in Brazil. I like to thank all the scientists and staff at RRIN who prayed and gave me moral support from the time Dr. Okaisabor selected me to the time I returned from Brazil to Nigeria.

Special thanks go to Dr. Richard Hall, Distinguished Professor of Forestry and Forest Genetics, and the Department of Forestry, Iowa State University who accepted my request for the one year deferment of the start of my Ph.D. program so that I could undertake the once in a lifetime germplasm expedition. In fact, Dr. Hall actually indicated that it would greatly enhance my ability to succeed in graduate school and eventually in my professional career. For this I am very grateful.

The ontogeny of this book as three chapters of a much larger book covering 25 years of international experiences occurred during my sabbatical leave in the spring of 2002 at the

Institute for Forest Genetics and Tree Breeding (IFGTB), University of Gottingen, Gottingen, Germany (UGGG). The sabbatical leave was funded through a study fellowship from the German Academic Faculty Exchange Services, Deutscher Akademisher Austavschdientst e.V. (DAAD). Dr. Reiner Finkeldey, Professor and Chair, IFGTB, UGGG, and Dr. Hans Hattemer, Retired Professor and Chair, IFGTB, UGGG, hosted me during the three months sabbatical leave at UGGG. Thus I am very grateful to DAAD, Dr. Finkeldey and Dr. Hattememer and, the faculty and staff of IFGTB for providing me an atmosphere to start the beginnings of an autobiography that has morphed into a book series. Special thanks go to President Humphries and Florida A&M University for granting me a sabbatical leave to enable me enhance my professional development as a faculty member of the university. While focusing on my primary disciplinary area during the sabbatical leave, it offered me the unique quiet moments to "put pen to paper" and hence the outcome of a book series.

Many thanks to Cynthia Barnett, a manuscript critique, who during the 2010 Tallahassee Festival of Books and Writers Conference organized by the Tallahassee Writers Association (TWA), suggested that I convert a manuscript that was already at 250 pages and expected to become 700 or more pages into a book series. 'Rubber Seed ----' is the first of the series. I like to thank members of my St. Louis Catholic Parish Renew Group, Mr. Bill Woolley, Mrs. Kay Woolley, Mr. Chris Hawkins and Mrs. Eileen Hawkins and of course my wife, Mrs. Lucy Onokpise for the tremendous spiritual support and encouragement they gave me as we meditated together on a fourth-nightly basis at each others residence. My gratitude and appreciation are also extended to Mrs. Hawkins, herself a published poet, who not only introduced me to my publisher she was kind enough to write the forword for this book.

The first four chapters of this book were read by my colleague, Professor Saleh Rahman, of the Institute of Public

Health at the College of Pharmacy and Pharmaceutical Sciences, Florida A&M University, while we were both traveling in Northern Ghana, on assignment from FAMU to the University for Developmental Studies (UDS), Tamale, Ghana. I learned a lot from him when I found out that as a physician and public health specialist he had already published several award winning literary books. That came as a great encouragement. I thank him for the emotional support and suggestions.

The completion of this book would not have been possible without the edits from many colleagues and students some of whom I contacted on short notice. I am extending deep appreciation to all of them: Dr. Olugbemi Moloye, Associate Professor of Anthropology, College of Behavioral and Human Sciences, Florida A&M University, Dr. Hank Bass, Professor of Genetics, the Department of Biological Sciences, Florida State University, Dr. Don Rockwood, Emeritus Professor of Forest Genetics, School of Forest Resources and Conservation, University of Florida, Dr. James Muchovej, Professor of Plant Pathology, College of Agriculture and Food Sciences (CAFS), Florida A&M University (FAMU), Dr. Alfredo Lorenzo, Professor of Landscape Design and Management, CAFS, FAMU, Dr. Clifford Louime, Research Assistant Professor, CAFS, FAMU, and Mr. Nandkumar Divate, Agricultural Development Specialist, Office of International Agricultural Programs (OIAP), CAFS, FAMU, as well as my graduate students, Ms. Lissa Reid and Ms. Angel Forde and an undergraduate student, Ms. Jade Walker.

Finally, I like to express my heartfelt gratitude to my publisher and writing mentor, Dr. Cheryl Jennings, for her patience, guidance and steadfastness in ensuring that we developed a great product known as Rubber Seed.

## PREFACE

In 1976, my director asked me to join a three-member team of two scientists and a lawyer to investigate the impact of gas flare on the macro and micro fauna and flora of an oil producing community in the Niger Delta. Little did I know that this was going to be the beginning of my international experience. This investigation exposed me for the first time to some of the activities of the oil exploration business. We undertook preliminary measurements of the temperatures leading from the pavement of a highway to the epicenter of the gas flare from an oil well deep in the forest of this Niger Delta Community. As we completed this investigation, I realized that we were actually investigating the most productive and largest multinational company, Shell Oil of Nigeria at the time. This of course marked my first encounter with an international body and the sowing of the seeds for international work. This exercise formed the springboard for a 30-yearlong international program activities.

Then came the selection to attend an international workshop on Hevea Breeding in Kuala Lumpur, Malaysia, in 1977. It represented my first foreign travel from Nigeria. I was in the midst of preparing for this assignment when my father suddenly died in a ghastly car accident that claimed a total of four lives. While traditional burials were completed for him, a formal church burial has never taken place to date. Encouraged by the director of the Rubber Research Institute of Nigeria (RRIN), Dr. Ayeke, who selected me to attend this workshop, I resumed the preparation and finally went to Malaysia. It was a therapeutic move to say the least, because while in Kuala Lumpur I focused on the breeding and genetics of *Hevea brasilensis.* It offered me an opportunity to interact with the many scientists, and in particular plant breeders who were carrying out breeding and selection work of natural rubber. Although returning to Nigeria brought back memories of my lost father, I had gained my first experience in traveling long distance by air. No doubt, this was a very long flight and it taught me what to expect in future international program activities.

Perhaps what I consider the most exciting and most rewarding of all my international activities to date came in early 1981. It was during the time when *ex situ* germplasm preservation of plant species was very popular. The initial concern for the rapid deforestation of the Brazilian rain forest was gathering momentum. It was very well documented that the Amazon basin and its related rain forests are the centers of origin/diversity of many of our cultured plant species. Natural rubber or *Hevea brasiliensis* happened to be one of them. Although Brazil has never been able to become a major world producer of natural rubber because of devastating South American leaf blight disease, it remains the repository of the gene pool for Hevea as its center of origin. From Rio de Janeiro to Brasilia to Manaus, from Cuiaba to Conomali/Cotriguacu, and from Aracatuba/Itauba to Villa Bella, we reached the very heart of the Brazilian rain forest. Working through estradas used for rubber tapping by the native Indians was indeed a very challenging, exciting and rewarding experience for me. It was the first time I flew in a single-engine plane when one had to enter the plane through the wings. It was a humbling experience for me. In some places we spent twelve days on a river boat during which we saw nothing but the waters of the river, the sky overhead and the thick wall of the Amazonian jungle on both sides of the river. I had a narrow escape of nearly stepping on one of those Amazonian orange colored snakes, but thanks to the quick action of one of our tree climbers who saved me from imminent catastrophe. Interestingly, after the four months in the Amazon Basin, we had to quarantine in a temperate country and so by special permission I spent a week in Canada, thus enabling me to meet with some old friends at the University of Guelph. Returning to RRIN and recounting these experiences to my family and then to colleagues and senior officials including my director, was indeed a milestone in the professional development of an African Scientist to the Amazon. Therefore, a major objective of writing this book thirty years after this tremendous international undertaken is to share with the reader the efforts often made but unrecognised by the public of what it takes to keep the industrial engine of the world running, and the great need to protect, conserve and continuously improve on those renewable natural resoursces we sometime take for granted. Recalling many of the activities in the jungle has been a great challenge but was quite refreshing and rewarding.

# FOREWORD

Plants, especially rubber seed plants are unknown to me. "***Rubber Seed: An African Scientist in the Amazon Basin***" is a page-turner. Kome Onokpise is an informed scientist whose story-telling style pulls you through the Amazon jungles, dodging choking vines and coiling pythons, as he scientifically outlines the purpose and importance of documenting the varied rubber seeds. It is evident during Dr. Onokpise's expedition why it is necessary to preserve this valuable natural element.

I am a poet, not a scientist. In this book learning why Dr. Onokpise is chosen by the Rubber Research Institute of Nigeria to participate in an international expedition to Brazil representing Nigeria and the International Rubber Research and Development Board in London is amazing. He is a quiet man. Seeing the unfamiliar words *"germplasm,"* the Amazon and rubber seeds hints at the adventures Dr. Onokpise and his expedition mates have before them.

Dr. Onokpise, in his understated way, alerts his family and co-workers to his four-month assignment. His writing reveals his subtle humor when organizing his life before leaving for Brazil and how he handles unexpected events in Brazil.

The international team members gather in Brasilia. The countries participating are China, Cote d'Ivoire, Indonesia, Malaysia, Thailand and Nigeria as well as Brazil. The scientists travel to Manaus, capital of the State of Amazonas to the Center for Rubber & Oil Palm Research to be introduced to the research process. Three locations have been chosen for research: Acre, Rhondonia and Mato Grosso. Dr. Onokpise's team travels to Mato Grosso.

Dr. Onokpise's descriptions of the people, places and plants are poetic. He infuses the pages of this book with three-dimensional locations and personalities. The Rio Negro's "confluence of dirty water with the deep blue" of the Amazon opens our eyes to the intermingling varieties of life to be observed with Dr. Onokpise and his teammates.

This book will enrich your life and open your eyes to more of the wonders in this world.

**Written by Eileen Sperl-Hawkins**
**Author of "*LOOK!*" and "*ANOTHER DOOR*".**
**Chair, Writers Group at Osher Lifelong Learning Institute at The Florida State University**

# PROLOGUE

When Charles Goodyear discovered the process of vulcanization, a chemical process which converts rubber into more durable polymers by the addition of sulfur, in 1836, the natural rubber tree, *Hevea brasiliensis* went from a plant species in the wild known only to the native Indians in the Amazon Basin to a fully domesticated and important industrial tree crop. The Indians called it " cahuchu" hence the word "Caoctchouc" in some literature. Until about 1913, Brazil was the world's major producer of natural rubber obtained mostly from rubber trees in their natural ecosystem. However the survival of the rubber seeds in Java, Indonesia in the late 1800s started the initial production of a few rubber seedling nurseries and plantations in Southeast Asia. This coupled with the prevalence of the endemic South American leaf blight (SALB) disease on natural rubber trees in Brazil, changed the landscape for natural rubber production in the world. Thus, while Goodyear and Firestone rubber companies and Henry Ford in the early 1900s made every effort to establish plantations in Brazil without success, the new rubber plantations in Indonesia and the rest of Southeast Asia were beginning to thrive. With the development of rubber research institutes in the region especially the Rubber Research Institute of Malaysia (RRIM) new clones were developed through conventional breeding and selection. Firestone established new plantations in Liberia, West Africa, to take advantage of some of these developments. Natural rubber plantations spread throughout the West African region in countries like Nigeria and Cameroon. Thus today, Southeast Asia produces nearly 90 %, and West Africa about 9 % with the remaining 1% coming from Brazil and other South American countries.

Many of the rubber clones in production came from those seeds collected by Sir Wickham in 1876 and went to Java through the Kew Gardens in Britain. Thus the narrow genetic base which resulted in the decreasing rate of increase in

performance of the rubber clones developed in the 1960s and 1970s led member institutes of the International Rubber Research and Development Board (IRRDB) to organize a germplasm collection expedition to the Amazon Basin, the center of origin of *Hevea brasiliensis*, in 1981 so as to widen the genetic base of the surviving Wickham seedlings. According to the IRRDB, planning and organization for the "Germplasm 81" activities started with a seminar in 1976 held in Jakarta, Indonesia, by the IRRDB. It was at this seminar that the principal requirements were determined, including the identification of the most promising collection sites in South America, agreements with host countries, the establishment of an intermediate quarantine station, and the establishment of a permanent germplasm center in one or two IRRDB countries. At its first "Workshop on International Collaboration in *Hevea* Breeding and the Collection and Establishment of Materials from the Neo Tropics" held in Kuala Lumpur, Malaysia, IRRDB participants came to the following conclusions. There should be a single expedition to the Amazon Basin to include several countries, and two reception sites (germplasm centers) should be established, one in Asia and the second in Africa. Materials collected would be divided on a 50-50 basis between IRRDB and the host country. The IRRDB germplasm and multiplication center would be in Kuala Lumpur, Malaysia and would receive 75 % of the 50 % IRRDB share of collected materials, while the second center, in Abidjan, Cote d'Ivoire, would receive 25 % of the collections. It was also agreed that scientific information from these centers in Brazil, Malaysia and Cote d'Ivoire would become readily available to IRRDB members. Furthermore, all IRRDB institutes that bore the financial costs of "Germplasm 81" would share in subsequent distribution of multiplied materials. The entire project was estimated to cost $1.7 million.
This book is a reflection of the experiences gained during a unique international expedition to the Amazon Basin, Brazil. This is particularly true of my team in Mato Groso. The

collection activities, the social interactions and the quarantine experience make for a great reflection on international activities involving a major international scientific organization, IRRDB and its member institutes. The preparations and effective implementation by various agencies of the Brazilian government is a testimony to the importance of this very important natural product to Brazil. It also reveals the great organizational skills of fellow Brazilian Scientiists. And although other smaller expeditions have since been made to the Amazon Basin, none has been as comprehensive as "Germplasm 81" Given the cost and logistics, it may be necessary to delineate an area in the Amazon rain forest for in situ natural rubber germplasm preservation. Existing gene banks in Africa, Asia and Europe will complement these efforts.

# CHAPTER ONE

## SELECTION AND PREPARATIONS FOR THE AMAZON BASIN EXPEDITION

It is 1980 and we are in the midst of another growing season. I had just returned from one of the field plots when the Head of Plant Breeding Division, Dr. Kako at RRIN invited me to his office. He told me that the director, Dr. Kaisal wanted to see us for an important discussion. Dr. Kako always held me in high esteem. He had been told that I developed the Plant Breeding Division, and established additional field plots of unevaluated clones. I had just started several trials in the field. On my part, I respected him for two reasons (1) he was as open minded as Dr. Yeke the former director of the institute, and (2) he recently transferred his service from the Nigerian Institute for Oil Palm Research (NIFOR). I read some of his papers earlier on in my career as we were building the Plant Breeding (now Crop Improvement and Biotechnology) Division. As we walked to the director's office, I wondered what could be so important that my divisional head could not tell me. The Director, Dr. Kaisal, had replaced Dr. Yeke over the past two years. Unlike Dr. Yeke, he maintained a heirechical approach to his administration. Yet he was quite considerate in his management style.

Upon arriving at the office, Dr. Kaisal warmly welcomed us into his office. He motioned us to sit down as he sat in a chair away from his regular desk. Thus, it became an informal setting.

"Good afternoon, gentlemen," Dr. Kaisal said.

"Good afternoon, sir," we both replied. Then Dr. Kaisal addressed me directly and said: "Mr. Onokpise, Dr. Kako and I have discussed the possibilities of you representing the Rubber Research Institute of Nigeria (RRIN) for germplasm collection in Brazil. The IRRDB has sent us an invitation which the Federal

Ministry of Science and Technology has approved." Continuing, Dr Kaisal said, "I have spoken to Dr. Kako about it. He has declined to participate on the basis of his age. Mr. Aka is not favored to go because of other commitments. Furthermore, I understood that you were the one who attended the *Hevea* Germplasm Workshop in Malaysia, in 1977. Thus, you are the one most familiar with the background to this exercise." Then he asked "Mr. Onokpise, are you prepared to visit Brazil??"

It was as if I was day-dreaming and for a few moments I could not say a word. I was pleasantly shocked by the comments of Dr. Kaisal and my selection to participate in an international expedition in, of all places, Brazil. Gathering my thoughts and not wanting to lose this unique opportunity, I said, "Thank you very much Doc. I am highly honored, for selection to participate in the germplasm collection. However, it will be very deeply appreciated if the director will allow me 24 hours to discuss the matter with my wife. She is pregnant with our second child and I am not sure how she will react to the news of my being away for four months."

Later that evening I spoke with my wife. I told her, "I have been selected by the director to represent RRIN and Nigeria in an expedition to Brazil. The expedition will last four months. My director has given me 24 hours to accept or reject the selection. I told him that I needed time to discuss the nomination with you before I could accept or reject the nomination" pausing before I finished the request. "Lucy, I feel this is a unique opportunity for us. It will greatly enhance my career." I was not prepared for what my wife had to say. 'Kome: You can not go!! We have just returned from Canada, I am seven months pregnant, and you are talking about being away for four months. Who is going to be around when I give birth to the baby? Also, remember our senior son is only four years old. How do you want me to manage the young ones and myself? No! No!! No!!! I don't think it is right for you to travel overseas at this time. You will have to reject the nomination regardless of how important this may be for your career now and in the future. You must put this family first."

To say that I was surprised by my wife's stern reaction as she walked back and forth in the living room would be an understatement. For a while, I sat stunned without saying a word. I thought she would be excited and welcome the news of a unique opportunity not available to many scientists in my field of specialization. Then as the thoughts went through my mind, it occurred to me that we were in June of 1980. True, we did return from Canada in May, a few weeks after defending my Master's thesis. Because the Institute (RRIN) had sent my family and I air tickets and had given me a specific date to return, I did not even participate in the graduation ceremonies. Suddenly, I remembered my director saying that the expedition would take place from January to April 1981. My wife's expected date of delivery was September 1980. My eyes popped and my confidence rose immediately. Still apprehensive, I got up from my seat and walked to my wife and asked her to calm down and relax, given her condition.

Thereafter, I said, "Lucy, we can have the delivery and still be able to go to Brazil. I just remembered that the director said the expedition would take place from January to April 1981. Our new child will be four months old before I travel. By that time my mother will be here from Lagos to assist with babysitting. Therefore, taking care of the children should not be a serious problem." Continuing, I asked her: "Could I tell my director yes, that my wife has approved the selection and hence I could go on the expedition?" The thirty seconds or so that I waited for my wife's response seemed more like three hours. She straightened up, looked at me, smiled and said, "Kome I am sorry I was so hard on you. I wish you had given me these dates when you made your earlier presentation." I was still very much in anticipation, not knowing whether the answer would be yes or no. She continued to make her intentions known but she was much calmer and spoke a little slower and finally the word I was waiting for was mentioned. "Yes, Kome," you can go on the expedition as long as your mother is here with us."

Then I embraced her profusely and thanked her while apologizing for not providing all the details at the beginning of our discussion. With my wife's approval I could respond to the director in the affirmative. The next day as soon as I got to the office, I went to Dr. Kako's office and told him that my wife was very supportive of my going to Brazil. I thanked him and requested that we seek an audience with the Director, Dr. Kaisal, as soon as he was available. Dr. Kako was thrilled and jokingly said, "All that estacode (per diem) you are going to receive, make sure I get the equivalent of one month's salary as my share." I laughed and replied, "let us wait and see how things turn out after our meeting with the director."

Dr. Kako then sent one of his messengers to confirm the availability of the director. About two hours later, Dr. Kako came to my office and said the director was now ready to see us. I dropped everything I was doing and joined him to see the director. "Good morning, gentlemen." He bellowed. "Mr. Onokpise, has your wife granted you permission to travel?" with a smile as though, it was not "manly" for husbands to seek advice from their wives when making such important decisions. Of course, this was Nigeria in 1980, the culture considered it submissive for a man to stoop so low as to request a woman's opinion before undertaking any major assignment in his life. We all laughed about the "permission" statement. "Yes, sir!" I replied. "After a difficult conversation, sir, she said it was okay for me to go." "Congratulations are in order." Dr. Kaisa said "Here are the application forms and other documentation that you should fill out and return to my office within 48 hours. These items are to be sent to The International Rubber Research and Development Board (IRRDB) in London. Upon approval, a formal letter of acceptance will be written to you that will enable you to obtain the necessary travel documents to Brazil." "Finally," he said "congratulations" again. We are counting on you to do a very good job. Dr. Kako has a lot of confidence in you and you must not disappoint him." Dr. Kako and I simultaneously thanked him. Dr, Kako reassured him that I would be an excellent RRIN representative. On my part, I thanked him again and promised him that ood morning "I shall do my very best during the international expedition effort" Dr. Kaisa then shook our hands and wished me the very best of luck in preparing and eventually participating in the exercise. It was as if I was scheduled to travel the next day when in fact there were still at least six months of preparation before the expedition itself. Of course, everyone was excited by the fact that the Rubber Research Institute of Nigeria was one of the institutions selected to participate in this unique international effort.

As we walked back to our offices filled with joy, I told Dr. Kako that I least expected to be selected for such an important exercise. Furthermore, my impression when I left for Canada in 1978 was that the germplasm collection would have been completed prior to my return in 1980. This must be God's work. Dr. Kako simply said, "You are very lucky, young man. This is going to open up other opportunities for you." We shook each other's hand, smiled and went into our separate offices.

Having been assured of the selection it was now time to focus on getting ready for the expedition. True, the expedition was six months away, but there was a lot to be done. My wife was pregnant, we had just returned from Canada and not much had been unpacked, nor was our residential apartment finalized. There were relatives anxious to see us. This meant my brothers and sisters, for whom I was responsible since the death of my father. They had not seen me in two years and were going to visit and spend time with me and my immediate family. I had to arrange with my mother to take care of her grandchildren. All of these required extreme care and planning skills.

At the workplace, we were initiating new research experiments including a major project of testing rubber clones at different locations throughout southern Nigeria. This was a study that I had initiated two years earlier, prior to my going to Canada for my Master's degree. A colleague of mine had ensured that the project concept was kept alive. Even the location of my office space had not been fully settled. I was actually using a temporary space as an office. There were no desks, air conditioning system, fans, and window drapes among other things. The laboratory needed significant improvement. With the new knowledge acquired from an industrialized country like Canada, my goal was to transform the Plant Breeding laboratory into a respectable laboratory with the limited resources available.

All of these thoughts went through my mind over the next few weeks. It suddenly occurred to me that there was a major item which I had forgotten. Prior to leaving Canada, I had applied and received admission to start my Ph.D. in Tree Breeding and Forest Genetics at Iowa State University, in the United States of America, with a start date of August 1980. I had also just been awarded a post-graduate scholarship by the Federal Government of Nigeria to undertake the Ph.D. program at Iowa State University. However, I did not receive the official letter of award until the latter part of June 1980, a month after returning from Canada. Because I did not want to lose the opportunity, my first order of business in preparing for the Brazilian expedition, was to quickly send an express mail letter to my designated advisor, Dr. Richard Hall. In the letter, I requested for a one-year deferment to the start of my Ph.D. program. Thus, instead of August 1980 I would appreciate it if my program started in August of 1981. I explained to him the unique opportunity of going to Brazil as a plant breeder on a germplasm collection exercise. His reply, one month later, was reassuring and very supportive. He indicated that the Department of Forestry in which I was going to obtain my Ph.D.

degree approved the one-year deferment. On a personal note, he wrote me, saying that many breeders and geneticists were always in search of opportunities to undertake such an international expedition and that the experience would add strength for my Ph.D. program. I was elated by his response, knowing that my admission was secured for another year.

For fear of not losing my Federal Scholarship I obtained three days of annual leave to visit the Federal Ministry of Education in Lagos. I took along with me three critical items: i. the official letter from Dr. Kaisal, my Director, selecting me to travel to Brazil, ii. my official letter of admission to Iowa State University, iii, the letter from Dr. Hall granting me a one year deferment for the start of my Ph.D. program. Upon arriving in Lagos, I spent the night with my sister, her husband and my mother who was in Lagos taking care of my sister's children. That evening, I told them the exciting news about my selection for a Brazilian expedition. They were all very pleased with the developments. My mother even mentioned that my late father would have been extremely proud of seeing his senior son undertaking such important assignment. The next day I met with officials of the Federal Ministry of Education regarding changes in the starting date of my Ph.D. program at Iowa State University. Fortunately for me I was directed to the Deputy Permanent Secretary of Education in charge of higher education and postgraduate scholarships. He was surprisingly very friendly and within two hours of arriving at the ministry I received an official letter signed by the Permanent Secretary approving the one year deferment so that my scholarship would take effect in August 1981 instead of 1980. I was very pleased and excited. I did not border to ask why they had been so friendly and treated me with such kindness and quick attention. I later found out that the official who handled my documents was from the Delta part of Bendel State. And that she was very pleased to see a young man from her part of the country making such progress. As soon as my documents were handed to me, I ran out of the building and took a taxicab to my sister's place in Ajegunle, Lagos. I told my mother, my sister and my brother in-law that because the Federal Ministry of Education gave me a very surprising and early attention coupled with the fact that it was still early in the day I could still get the Bendel Line to Benin City on the same day. There was no need to spend another night with them especially as my wife was alone pregnant and managing our senior son by herself. I did however remind them that my wife and I were expecting my mother to come to Benin City once my second son was born.

When I got to the Bendel Line station the next scheduled bus was loading on the passengers. I immediately ran into the office and bought a ticket. With no luggage, there was no problem securing a place in the bus. There were no major events on the journey back to Benin City. The next day, I resumed work a day earlier than I had expected. I was now highly motivated to work and my wife, was already gearing up for the arrival of our baby in a few months. That afternoon, Dr. Kako stopped by my office and noticed that I was already at work. "Mr. Onokpise, I thought you took annual leave to travel to Lagos to straighten your Federal Scholarship?" He asked after our usual traditional greetings. "Yes, Doc. I received a very surprise early attention at the ministry. It was an unexpected positive outcome, so I decided to return home early." I replied. "But you know, you could have stayed home today, relaxing and taking care of your family, and then resume work tomorrow." He said, smiling. "Thank you very much, Doc, for your concern. I really appreciate it. I am trying to get as much work done before my wife delivers our baby when I will be away for a month's annual leave. There will be very little time left when I come back from that leave and getting ready for my Brazilian trip." I responded with a deep sense of commitment to an organization that had just elevated me to unexpected heights in my very young professional career. He nodded, looked at me, smiled and went to his office.

Meanwhile, my wife started early preparations for the baby's arrival. She arranged with one of her younger sisters to spend more time with us. Since her family lived in Benin City as well, it was relatively easy for her siblings to come spend some time with us quite frequently. All the baby items had been purchased as we waited for the expected day of delivery (edd). Some of her old classmates who heard we had returned from Canada also paid her visits. On September 10, 1980 my wife delivered our second child at the St. Philomena's Catholic Hospital in Benin City. I had gone to work early that morning not knowing whether Lucy was going to give birth. The only indications were the very excruciating pain she experienced throughout the night. The pains had dissipated just before 5:00 a.m. after taking some medication. So I felt confident to go to work. One of my younger brothers, Christopher, was on holidays and leaving with us along with one of my sister's in-law, Clara. I had asked them to do whatever my wife asked them to do until I got back from work. As she told me later, at about 10:00 a.m. that morning those pains resumed, so she quickly got everything ready and asked Clara and Christopher to get a taxicab which they took to St. Philomena's Hospital. And just two hours after arriving at the hospital she gave birth. It was very interesting to note that, this was the same hospital she had been born some twenty-seven years earlier. She graduated from this same hospital with a midwifery diploma in 1976.

I went to the hospital straight from work as a colleague of mine who was at the same hospital at about the same time with his wife when my went in to the maternity area informed me that he saw my wife at St. Philomena. I was overwhelmed when I walked into the private room where Lucy and our new baby were. My mother in-law had also come to help as news reached her of another grandson being born by her senior daughter. The next day I sent a telegram message to my sister in Lagos announcing the arrival of my second son and another grandchild for our mother. In the telegram I reminded my sister of the promise she made to allow our mother bring her two children to Benin City where she could babysit all four grandchildren at the same time. It was also a commitment I made to my wife, Lucy, so that she could approve my selection for the Brazilian expedition

Within three days, my wife was discharged and she came home with an eight-pound bouncing baby boy. We arranged a welcome party that was very well attended. The neighbors and my colleagues from RRIN came to welcome our new baby. Three weeks later my mother arrived from Lagos. Her arrival was very timely, because it freed me up to undertake other shores and focus on returning to work after nearly one month of annual leave. Of course with four young grandchildren, ages one month to 5 years, my mother's hands were full to say the least. You could see the joy and smile on her face as she went about caring for the children as if it was nothing to be concerned about. With the arrival of our baby boy, a major hurdle for the Brazilian trip had now been cleared. I returned to work on October 3, 1980 after nearly one month of annual leave.

In mid-October the letters of approval of my selection for the international expedition arrived from IRRDB headquarters in London. There were so many forms to complete in order to obtain travel documents. Having just returned from Canada I did not need a new passport since my passport was current, but I had to complete the forms for a visa to Brazil, security checks and others. A gentleman, Olomu, working in my director's office developed a not-so-friendly attitude regarding the processing of these documents. I could not understand why he behaved this way towards me on this particular issue. He seemed to have been very nice when I first returned from Canada. On other matters he had no problems processing my paperwork on–time. However, when it came to the Brazilian expedition Olomu delayed everything to the last minute. I could not figure out what the problem really was. On several occasions I asked him why he was so uncooperative when it came to processing my papers for Brazil. He never did respond to my questions. It was only after I had obtained my visa to Brazil in October of 1980 that someone told me that Olomu who was a friend to Mr. Ali, felt that I was the youngest scientist in the Plant Breeding Division and that I had just returned from Canada, therefore if Dr. Kako was not going on the expedition then Mr. Ali should have been the next person in line to be selected. At this junction, it occurred to me that a few years earlier, Olomu was opposed to me going to Canada for the Master's degree. Hence I should have known better, but in the few months since I returned from Canada, he had been one of the most pleasant workers in the director's office. Hence I did not pay much attention to his usual antics.

The Brazilian embassy invited me in late October 1980 for a visa interview. This meant, I had to make another trip to Lagos. I have never enjoyed staying in Lagos even though I attended the Federal School of Science, Lagos from 1969 to 1971. As usual, I stayed with my sister and her husband in Apapa. We spent the evening having a very nice dinner of yams, plantains, palm oil and the ubiquitous pepper soup, a delicacy in its own right. My sister, Joe, asked about our mother, her children and mine as to her our mom was coping. She promised to visit as soon as she got one of those long weekend off. She was very eager to see my new son. In fact she had neither visited us nor had an extensive discussions with me since my family returned from Canada. Yet, things were moving so fast. "Kome, here you are in Lagos again in just five weeks." She clapped her hands as she sipped a spoon of the "hot" pepper soup. After dinner we continued our conversation late into the night recalling our childhood times in Lokoja and how far we had come with God's blessings. Only 18 months separated Joe and me in terms of age. Although my junior, we had developed a very healthy sibling rivalry. She was often referred to as our "father's girl" and I as our "mother's boy." No doubt, the fact that I was raised by our maternal grandparents, the late Mr. and Mrs. Osuwe Agbanobi, until I was seven years old before rejoining my parents, while Joe was always with them could have played a major role in this natural outcome. Yet, over the years Joe and I had a strong mutual understanding and even when we had moved on into our professional and family lives, the respect for each other, as close siblings remained very strong. It was my brother in-law, Ernest, who came into the dinning area to remind us that I had an interview to attend and that he had to go to sleep because he needed to wake up to beat the horrendous Lagos traffic the next morning

As for the interview itself, it was a very interesting one. Whenever I had an official visit to Lagos I always wanted to be done very quickly and leave the city. So I was very pleased to see the interview completed within two hours of its commencement. The interviewing officer wanted to know if I spoke any Portuguese. I told him I could not speak Portuguese. Then he joked, "How are you going to do your job in Brazil without speaking Portuguese. Then I told him that the IRRDB and the Brazilian government had arranged to provide translators to the expedition team. After several probing questions regarding my background and other professional activities he smiled, shook my hands, and asked me to remain in the waiting room while the processing of my visa was completed. Shortly thereafter, I received my passport with an approved multiple entry visa to stay in Brazil for six months.

Elated, I ran out of the Brazilian embassy to catch a taxi to Bendel Line Transport for my return trip to Benin City. Bendel Line was one of the most reliable (if not the most reliable) means of transport from Lagos to Benin City. It was operated as a para-statal or a quasi-government enterprise. Also, it was relatively inexpensive compared to similar bus lines at the time. The journey from Lagos to Benin City was quite smooth given the nature of the roads. Furthermore, because I was so tired from all the activities, I slept through most of the journey that lasted four hours. Upon arrival, I took a taxi to our flat, because we had no car, having sold the one I had before I left for Canada. Our hope was that once I settled down, my wife and I would consider buying a new car. As I entered the flat, my mother and my wife welcomed me with broad smiles. I asked about our newly born baby. Already, my senior son who had just turned four years old ran from his room where he was playing to welcome me. "Papa, come and see my small brother in his crib." Maro, my senior son said. "Grandma has been helping mama to take care of Omurhu, the name of our new baby boy." He continued playfully. There was so much excitement in the house, I felt very blessed. I hugged my wife thanking her for the wonderful job she was doing raising the family. Then I went to the crib to pick-up Omurhu who by now had woken up. I came back into the living room with him and Maro.

My mother who was in the living room, surprisingly watching a TV program, looked at me, and her two grandsons with such a broad smile. She said she felt a sense of satisfaction at watching my progress in life and raising her own grand children. As we sat down to dinner, my wife asked me how my trip to the Brazilian Embassy in Lagos went. I told her that everything went very well. There were some questions, including the one about being able to speak Portuguese. She laughed and said, "Perhaps, you can use your understanding of French to help you." "Thank you very much, Lucy," I replied.

The next day I took my passport to work and went to see Dr. Kaisal to show him the visa. "Sir," I said, after the usual greetings. "I have obtained my entry visa to Brazil." "Very good," he said and "congratulations." "All you need is your air ticket and you will be ready to go." Then with a sarcastic smile he asked, "By the way, has your wife seen the visa?" I replied smiling, "Yes, sir! She saw it as soon as I returned from Lagos." I thanked him again for the opportunity and promised to do my best. I left his office smiling all the way.

# CHAPTER TWO

## DEPARTURE FROM LAGOS AND ARRIVAL IN BRASILIA

My flight was scheduled to go directly from Lagos to Rio de Janeiro. I flew on the Brazilian Airline, VARIG International. This was January 1981. The Nigerian economy was strong and the socio-cultural system of the country was very stable. Both Brazil and Nigeria as leading third world nations had established cultural and economic ties. As a major oil producer, Nigeria supplied oil directly to Brazil in exchange for manufactured goods. These were the days before the evolution of the G-20 countries. Thus the air link to Rio was a natural outcome of this growing economic relationship. It was the time when the movement for a south-south trade was gathering momentum. Thus, a flight that would have taken over eighteen hours lasted only six hours across the southern Atlantic Ocean. The VARIG Flight attendants were extremely pleasant. They went about the cabin with a lot of smiles and responded to all requests with politeness and dignity.

This was my first flight to South America and my third flight overseas. I had become more confident and enjoyed flying. Also, it was an era in which airline services were at their best. The hostesses were very courteous and responsive to passenger's requests. Fewer people flew at this time and so there was respect and cordiality among passengers. It was an enjoyable flight and we touched-down safely in Rio de Janeiro. Immediately I joined the connecting flight to Brasilia. This flight lasted about forty-five to sixty minutes according to my estimation.

As we started going through the immigration process, I suddenly realized that I was going to have a language problem.

During the six months of preparation in Nigeria it never occurred to me or anyone for that matter that the official language of Brazil was Portuguese. I was told that our host personnel would be fluent in English. I inadvertently ignored the Brazilian officer at the Brazilian Embassy who asked me about speaking Portuguese. I never expected to meet one of these officials at the airport in Brasilia. It turned out that I had arrived a day earlier than the international expedition team was expected to arrive. Therefore, I found myself using sign language for expression. But not being so fluent in sign language, processing my travel documents and getting out of the airport became bogged down and a very tense atmosphere immediately developed. Here was a young African male trying to get into Brazil in 1981 to conduct an international scientific work without the necessary documents. Questions were asked.

Then a lady checking through some baggage several meters away from the passport and health sections beckoned for me to come over to the baggage checking area. I noticed one of the men speaking to her and within a short time, she moved towards my direction. I was feeling disturbed and a little afraid as she approached me wondering whether my travel documents were not in order. If so what was going to happen to me. I dreaded the idea of being asked to return to Nigeria after the excitement that my selection and departure for this expedition had generated.

Then she smiled as she stopped in front of me. "Parlez-vous francais?" In a state of shock for a moment I could not immediately respond. I realized that this was a God sent person, and that I had studied French for five years in my secondary school, looking forward to the day that I could use it. I recalled my wife's joke about "using that your knowledge of French ------." Little did I know that the joke would become a reality so soon upon arriving in Brazil. I gathered myself, smiled and responded in the most polite manner my parents taught me to respond to a government official on duty.

"Oui, Je parle un peu!"

The relief that filled the air could be felt for miles away. All the other officials could be seen relaxing and frowning faces started to show smiles and eyes opened up and a friendly atmosphere prevailed wondering who this young African was. It became obvious that these immigration officials had not seen many Africans traveling to Brazil to be guests of the Brazilian government.

She guided me through the immigration lines and onto the baggage claim and checks while conversing with me in French. Just as we moved out of the baggage area, I saw a man waving in our direction. My newly found friend increased her steps and as we reached him, she embraced and kissed him with a broad smile. Since they spoke in Portuguese I could not understand what they were saying. Shortly, after, the gentleman stretched out his hand towards me for a handshake and he said,

"Soyez le bienvenu au Brazil."

"Merci beaucoup." I replied.

The couple picked up my luggage and escorted me to their car. Then we drove towards the Bristol Hotel which was listed in my documents as to where all the scientists would assemble in Brasilia. As we drove from the airport to the hotel we conversed in French. The couple wanted to know where I studied French to be so fluent. I told them that in my secondary school we had to study French and Fine Art in the first four years and in the final year we had the option of taking French or Fine Art. I was one of only four students who chose the offer of French for the West African School Certificate (WASC) at the time. And because it was mandatory for us to pass the oral part of the French exam in order to be successful in WASC, I had a lot of practice speaking French in order to succeed. I also asked them how they came about being so fluent in French. My new friend said that her husband used to be a member of the diplomatic corps in France as an employee of the Brazilian

Foreign Ministry. They spent a total of five years in Paris. When she finished, I said, "that is very nice. I would like to visit Paris someday." "OH!" she replied. "Paris is a very beautiful city and the people are very courteous. I strongly recommend that you visit there whenever you can."

At this point, her husband pulled into Hotel Bristol. They assisted me with my luggage and then helped me into the check-in counter. Then as soon as one of the hostesses spoke in English, my "friend" smiled as if to say "you are now in good hands and we can now go home." Her husband shook my hands and wished me the very best throughout my stay in Brazil. I followed them back to their car and as they pulled-off I waved to them with a very broad smile in my face.

I returned to the check-in counter, collected the keys to my assigned room and took the elevator to my room. As soon as I opened the door, I put down my luggage and sat down for about thirty minutes to reflect on what had transpired at the airport and how the knowledge of a second international language had rescued me from a sticky situation. I also wondered how God was so kind to have a Brazilian at the airport who could speak French and a couple that was so nice to a foreigner visiting Brazil for the first time.

On the second day at the hotel, the actual date for all the international scientists to arrive in Brazil I observed that some notices had been put in place to welcome IRRDB members. Suddenly, I realized how lucky I was to have been brought to the very hotel where my anticipated teammates were to be housed in Brasilia. I spent the early part of the day walking around the hotel vicinity to view Brasilia and to reflect on the task ahead. At 29 years of age, here I was in a foreign country involved in a major international scientific expedition. It was a dream-come true. I felt very proud of myself, my parents, my young family and the Rubber Research Institute of Nigeria (RRIN), but also realized the daunting task ahead of effectively representing my research institute. As I closed the door to my hotel room and settled down to review the documents in my brief case, a calm feeling came over me. "Just think I am part of the largest international scientific team to undertake the germplasm collection of one of the world's major industrial crops."

Early the next morning, members of the international team assembled for breakfast. At this time the team leader for the IRRDB scientists, Dr. Subramanyam and his Brazilian counterpart, Dr. Paulo, who was actually the overall expedition leader for the germplasm collection exercise, welcomed all the scientists informally. They indicated that a formal meeting had been planned for 9:00 am. at the conference room of the hotel for briefing. Once in the briefing room I noticed that there were scientists from China, Cote d'Ivoire, Indonesia, Malaysia, Nigeria (myself) and Thailand. We were joined by scientists, from the Brazilian National Center for Genetic Resources (CENARGEN), under the Brazilian Enterprise for Agricultural Research (EMBRAPA). I recall that the Brazilian Minister for Science and Technology welcomed us to Brazil and commended us for undertaking the important task of preserving and conserving the gene pool of one of the world's most important industrial crops. The Director of CENARGEN then spoke, stressing the importance of genetic resources especially of natural rubber because its center of origin was in Brazil. Following three hours of presentation by various officials we had a lunch break. The afternoon session lasted another three hours concluding with plans to visit major institutions in the capital city Brasilia. At a dinner held that night we met a number of high level dignitaries from Brazil and the various countries represented on the IRRDB scientific team.

As I settled down for the night, I turned on the TV to listen to the late night news. To my amazement, I saw pictures of our program from earlier in the day. Although, I did not understand the Portuguese language, I concluded that the germplasm team expedition into the Amazon Basin was being relayed on the Brazilian TV. I observed that the TV camera was focused on each IRRDB member for 3 - 5 seconds. Thus, I saw myself on TV in Brazil doing scientific work. This was indeed overwhelming for me. As I contemplated this thought I turned-off the television and fell asleep. Over the next few days we visited the various institutions as planned for us by the officials of EMBRAPA. Meanwhile, all documents needed for us to work in the various states of Brazil were finalized. The tour of the city of Brasilia was very exhilarating. Looking back, and having traveled to many other parts of the world, I consider Brasilia one of the most well planned cities in the world. When I asked my new Brazilian friend how come the city is so well planned. His response was that "Brasilia was built from scratch and the Brazilians decided to move their capital from Rio de Janeiro on the seacoast to the middle of the country. In so doing it became necessary to involve architects, urban planners, engineers of all stripes and other professionals to develop and build a capital city that was befitting of the largest country in South America." "This must have been a very expensive project." I replied.

"Yes, of course" he said. As to the expense and source of such large monies my friend indicated that he had no ideas.

After about seven days in Brasilia we (the IRRDB scientists), and some Brazilian scientists from CENARGEN were flown to Manaus, the capital of the state of Amazonas. Here, we were hosed by the Center for Rubber and Oil Palm Research (CNPD). It was at this Center that the IRRDB scientists met their Brazilian counterparts, both of whom exchanged greetings, business cards and engaged in discussions surrounding natural rubber and the future of the industry. At a special session the next day, the overall leader of the IRRDB-Brazilian expedition team, Dr. Paiza del Rio gave a detailed briefing of the collection exercise. Dr. Rio was a very well respected Brazilian Tree Breeder and Geneticist. He spoke about the importance of the germplasm expedition to Brazil, and that was why the Brazilian

Government had invested so much man power and financial commitment to the project. He indicated that every few years Brazilian Rubber Tree Breeders and Geneticists collect a few germplasm but has not been sufficient to develop big plantations like many of the IRRDB countries. Continuing, he said "I may not be going on the expedition with you, but you have capable Brazilian scientists with you." He also described the geography, the people and what to expect in the Amazon jungle. He indicated that the entire operation for "Germplasm 81" within Brazil was coordinated by CENARGEN, EMBRAPA, an organizing committee made up of the Superintendent of SUDHEVEA, two Brazilian scientists and an IRRDB representative. Additional coordination at the state and local levels came from units of EMRAPA, the Brazilian Agricultural Extension Service (EMBRAPTER) and other SUDHEVEA Units. Items assembled for making the trip into the jungle, included material for both scientific work (vials and bark sampling equipment) and physical survival in the jungle (pocket-size jungle knives, maps, compass and water cans). A backpack weighing about 32 kilograms (70 lbs) for each team member was developed.

The team was given a tour of the Institutes of Research for Rubber and Oil Palm respectively. An official visit was made to the historical places in Manaus. The state of Amazonas as the name suggests is in the heart of the Northern Region of Brazil where the Amazon River traverses. And Manaus is the city where the Amazon and its largest tributary, the Rio Negro, have their confluence. Our host took the team in a ferry-like boat to this confluence. It was a very illuminating site. We could see the rather large dirty brown waters of Rio Negro and the dark blue waters of the Amazon River. At the confluence point, the body of water was very rough. This body of water covering over one square mile did not surprise me. Only its expanse and continued rough look. As an elementary age boy, I grew up in the town of Lokoja, the confluence town of Rivers Niger and

Benue. My father was an officer in the Inland Waterways Department and served as the Patrol Officer for many of the ships that plied the Niger and Benue Rivers. On several occasions I traveled with him and many times we went through the point where both rivers actually met. As a youth I was always fascinated by this wonderful sight. So, seeing the Amazon River and the Rio Negro at their confluence point brought back fun memories. To see this in Brazil was even more special.

After one week of orientation in Manaus, the three teams for three separate states were sent to the individual states: Acre, Rhondonia and Mato Grosso (MG) for germplasm collection. My team was sent to MG. Mato Grosso is located in the Central West Region of Brazil and lies to the south of the Amazonas in what is considered a transition zone between dense forests of the Amazonas and the Cerrado or wooded savannas in Southwestern Brazil. Hence, sites such as Conomali, had very thick jungle like equatorial rainforest, whereas, areas like Aracatuba had wooded savanna, open grassland areas. Thus in places there was cattle grazing and in other places heavy logging took place. The state capital, Cuiba, became our primary base camp. Contact was established with officials of SUDHEVEA who provided much of the logistical support needed for germplasm collection in the jungle. Flights into the jungle were initially conducted by an Aero-Taxi company but taken over by the EMATER planes soon after.

At the secondary base camp, often referred to as the serringal by the MG team, information on the availability of natural rubber material was sought by my team leader (Paez de Gasol). Depending on how early we arrived at the serringal, collection could start on the same day, otherwise information was assembled and/or we waited to get into the jungle the next morning. Food and other items were bought from the store in the serringal for journeys that lasted more than one day from our secondary base camp. Also, tour guides were provided in each serringal. Such guides had knowledge of the coloaccoa that was visited and an understanding of the seringueros themselves. On boat trips and very long road trips (done by trekking), an additional local person was added to the team as a cook. Thus, in the jungle, there could be eight or nine persons on the trail depending on whether we were collecting materials or moving from one location to the other.

In each estrada, to collect seeds, the team usually picked up any seed in site as we moved along the Estrada since the number of rubber trees in the estrada varied from as low as 5 trees to as high as 700 trees. Sometimes we stopped in an estrada and covered about 100 square meters to pick up the seeds and then return to the estrada to continue working on the trail. An estrada with many trees was divided into two so that seeds from the first part constituted one sample and seeds from the second part constituted another sample. This division was multiplied even more, when the team observed considerable amount of variability among the seeds. Germinated and broken seeds were not picked. The seeds were put directly into perforated polythene bags and a small tag placed in the bags for identification purposes. Occasionally, seeds were obtained from dried pods by cracking such pods with a wooden mallet. At the secondary base camp, the seeds were counted, reselected and placed in fresh perforated polythene bags, assigned one of the collection numbers and then made ready for dispatch to Manaus.

Because budwood materials had to be obtained from elite clones they were sometimes very difficult to collect. Such trees could be very large and so difficult to climb. At other times the elite tree may be at the end of an estrada or even in the middle of a flooded area. In any case, to get budwood from a given tree, the materos (tree climbers) climbed directly to the top of the tree if it was a small or an average size tree (0.5 to 1m in girth). If it was a very large tree (1.5m 4.0m in girth) the matero used an accessory route by first climbing an adjacent smaller tree. He was then pulled by means of a rope to the top of the large tree. Branches of the trees were cut. On the ground, these branches were examined and budwood materials were cut (each piece of cut branch being about 1 m long).

# CHAPTER THREE

## INSIDE THE JUNGLE:
## I. CONOMALI, COTRIGUACU AND RIO JURUENA

Conomali, our first secondary base camp in the State of Mato Grosso was one of the most exciting places for me during the expedition. It is located in the northern part of the state on the banks of the Rio Juruena. In 1981, this area was essentially a primary jungle, typical of a tropical rain forest. We went to Conomali by light aircraft, single engine plane from Cuiaba in about three hours. Up to that point my flights had been between the big cities in Brazil. These were from Rio de Janeiro to Brasilia, to Manaus and then to Cuiaba. So, the flights were on jet planes. Flying from Cuiaba to Conomali was quite different. I recalled that we had been warned during the briefings in Brasilia and Manaus that the only way to get around in many parts of Brazil at the time would either be through flying or by boat. It never occurred to me that such flying would be in small airplanes. Here I was entering a twin-engine plane large enough to carry only six persons. To get on the plane I climbed across one of the wings. As I strapped the safety belts around my waist, I noticed how close we were to the pilot and its operating instruments. As the plane took off, it was obvious that my expedition into the Amazon basin had begun.

As we flew over the carpet of trees, I noticed that this twin-engine plane was not too high above the trees and it kept close to a river whenever we came across one. The view below was exhilarating as each group of trees receded behind us. Just as I was getting over my fears and initial shock of flying in such a small plane, the pilot announced that we would be landing within twenty minutes. It was a breath-taking experience. As we approached the landing site I noticed that there were no buildings nearby and so I asked myself how and where it was going to land. Then a wide-open space appeared immediately below us and the pilot landed the plane without any problem. No doubt, this airstrip or landing strip showed evidence of regular use. One could see that even though located in the middle of "nowhere' the landing strip was well maintained and the pilot showed an excellent familiarity with the airstrip. Two young men came to the plane and off-loaded our luggage, which they carried to an area about 100 meters from the landing strips. Looking further, I saw a group of houses all of which looked more like log homes than brick homes. We were taken to the largest building in the area of eight to ten buildings overlooking Rio Juruena and greeted by a heavily bearded man in his mid-50s.

After 15 minutes of discussion I realized that this man was the patriarchal leader of the group in the place we called Conomali. The living room had interesting objects ranging from dried eagle and hawk feathers to dried whole Brazilian panthers. In fact, this man whom I referred to as George had a live panther as his pet. There were also dried skins of pythons and other animals in this enclave. George stroke his beard as he leaned back in his recliner chair with a sense of fulfillment seeing so many "Engineer d'Agronome" in his unique domain. This is not a place where people visited all the time especially international visitors. The smile on his face showed, as our team leader introduced each one of us to George. Following a warm welcome in which sodas (soft drinks) were served, we left this main building to smaller huts where younger individuals lived. I observed that bunk wooden beds had been brought to accommodate as many as four persons in a room. I surveyed the room to see how these men managed to survive in this deep jungle of the Amazon basin. As I made my observations I heard noises coming from a corner of the room. I asked our young host, Kevin, what the equipment was. He said it was a ham radio, which they used for communicating with the outside world. The ham radio also served an important function of guiding small planes like the one we flew in, to safe landing at the small airstrip in the area. Kevin then explained to me that the ham radio was battery operated, and without this ham radio they would lose complete touch with the rest of the world. Kevin became my very good friend over the next eight days at this secondary base camp.

Later that evening, we visited the banks of the Rio Juruena where the canoes and motor-powered boats were anchored. About 5:00pm that evening our team leader and three of us IRRDB scientists boarded one of the engine boats and rode across to the east side of the river with Kevin serving as the driver and guide. Some of the locals had mentioned to George in the past that there could be natural rubber trees in the area. So what we did that evening was to scout the area in preparation for an actual collection exercise. Dinner was served in the dining room of George's main building. We all gathered in his living area and the cook made large pots of rice and fresh steamed fish. The dinner was very delicious and the fish was very tasty. As someone who grew up close to a major river and whose father had to have at least one meal a day prepared with fish, I had a good feel for fish that had been caught directly from a river. I slept very well that night because I was very tired and the meal was quite refreshing.

Early the next morning, we were woken up by the sound of a bell from the middle of this huge compound. The gardeners were all out, the cooks were already cooking and our tour guide already arrived. By 6:30 am we were loaded into an engine boat with all of our germplasm collection materials and pulled to the site that we scouted the previous evening. With machetes and other small knives our tour guide and three climbers guided us to our first natural rubber tree. When I picked up my first official rubber seed in this thick jungle where you could see nothing but water, forest walls and the sky, I could not hide my youthful scientific emotion for "Germplasm 81" in the home of *Hevea brasiliensis,* I screamed with a broad smile. "This is RUBBER SEED in its original residence!!!" Everyone turned around and looked at me wondering what was going on. I told them about my inner joy to finally be engaged in the expedition in the world's most referenced and largest Tropical Rain Forest. Our team leader patted me in the back and our collection continued. We walked for another 15-20 kilometers before returning to the engine boat for lunch and continued downstream repeating the process performed on the first stop. At about 6:30 pm we completed our prospection for the day. We got back into the engine boat and sailed upstream back to our secondary base camp in Conomali. Again, we had dinner with our bearded host. On this particular night, I noticed that he was in a very light mood. He joked a lot and asked every one of us about our backgrounds and how our first day of germplasm collection went. Our Brazilian team leader led most of the discussion. Kevin who was quite fluent in English did all of our translation. As we finished dinner and settled down into the chairs in the living room, I jokingly asked George how he managed to come to Conomali. It was as though George had been waiting for this particular question to be asked.

Stroking his beard and with a grin on his face, George went on a long monolog with an occasional pause. His eyes opened wide at intervals, and shook his head and bowed down, any time he called on names he remembered as a colleague, friend or family member. According to George, he had served in the Greek infantry division just before the Second World War. Then after the Axis forces overran Greece he realized he could no longer serve in the Greek army, so he deserted. Rather than stay in Greece as a deserter in the middle of the war knowing fully well that the army of Hitler or Mussolini could have drafted or conscripted him, he found his way to Brazil. I asked him why he didn't settle into Rio de Janeiro or Manaus. He starred at me for a long time as if to say, "Young man, you don't understand". Then he went on to say that he had to get as far away as possible where he could not be reached by the Greek Government for deserting the army at a time when all young men in Greece were needed for liberation of the "mother land." You could see some fright in his eyes as he looked around the room to be sure that he was among friends and that none of us "so called scientists," was an agent of the Brazilian or Greek government. Sensing this obvious concern, I quickly told him that he had done very well for himself and that he had basically established a whole "town" in the middle of a dense tropical rainforest. I wondered how he was able to survive some of the most dangerous animals, snakes and fish in the world. He laughed loudly and said. "I was a young man then and I had just left the Greek army which was a very disciplined force at the time. So when I arrived in Brazil, I gathered enough money and materials to move out here. I directed my energies towards dealing with the Brazilian panthers, the python and other snakes, the piranhas or "man eating" fish as I built up this place. In some cases, I even became a friend to these animals." I was very fascinated by this unique story and wondered how a young man in his late twenties or early thirties could leave such a solitary live in this thick jungle.

At this point, I noticed that our Brazilian team leader checked his watch and reminded everyone that we needed to be up by 5:00 am the next day. I thanked George immensely for answering my questions. He shook my hands and patted me on the back and said that he never knew that Engineer d'Agronome could ask such questions. In Brazil, agronomists and plant breeders were called Engineer d'Agronome and were highly respected. He waved good night to us and we went to bed and I slept very soundly so that I was very refreshed when I woke up the next morning.

On the second day, we went through the same routine as the previous day except that we went further south downstream on the Rio Juruena. By now, I had started to observe that at every stop where natural rubber seeds were found, they varied significantly from the prior location either by patterns on the outside of the seed, their shape or on the color of the seed *per se.* I had to keep this in mind as the prospection for natural rubber seeds continued. This genetic variation was distinct in areas around river banks but not so much in upland areas throughout our collection exercise. It is possible that in flooded areas water helped to transport the seeds to longer distances under what is considered hydrochorous seed dispersal, leading to many ecotypes. Furthermore, the simple explosive mechanism of seed dispersal in upland areas, tended to limit the extent to which seeds could be dispersed. Unless such seeds were carried for long distances by animals (zoochorous dispersal), rubber seeds in upland areas get dispersed over shorter distances. On this day, we spent only eight hours prospecting for seeds, and returned to Conomali in the afternoon. After making my diary notes and remarks, I took a nap or siesta. When I woke up, I saw Kevin receiving and sending messages over the ham radio. While still lying on my bed, I asked him how much time he spent daily on radio messages.

"About two to three hours on a day, depending on the situation. If there are any emergencies in and around the area, then I stay longer otherwise I just listen and tell George what is going on in the area."

As soon as he finished with the radio messages, I told him that it had been three days since we arrived in Conomali and had not walked around this settlement. I pleaded with him to give me a quick tour. As you might expect we finished our tour of the settlement within thirty minutes. However, on one of the areas near the cassava farm, there were very many black polythene bags containing young cacao plants. This brought back memories of clonal nursery. I asked Kevin if anyone had thought of transplanting the plants to a permanent field. Kevin indicated that as long as he knows, no one had given thought to transplanting the plants. Perhaps nobody in the area knew how to transplant the seedlings and/or other materials partly because no one in the settlement knew the exact spacing for this type of tree crops. Then I asked Kevin if there was anything wrong with my arranging the gardeners to transplant the seedlings before my eventual return to Cuiaba. He looked at me, shook his head and started walking towards our building. Then I quickly caught up with him and wanted to know why he reacted the way he did. "You must be out of your mind." He said. "Those plants have been there for almost 30 months and everyone thinks there may be some dangerous snakes hiding among the polythene bags. Also, how long will it take for cocoa trees to reach harvest age and for us to enjoy the cocoa beans" he concluded. I said to him "If we transplant now, given the age of these plant and the fact that we are going to transplant them into a highly fertile soil, because not much farming has been done around here, the trees could be harvested in three to five years. And of course if you dry the beans properly and have them well processed you could enjoy your chocolate right here deep in the Amazon jungle."

As we conversed, Kevin was walking towards the bank of the Rio Juruena and picking up one of the canoes and fishing hooks. Another scientist saw us and came down and very soon there were five of us in the canoe. As we pulled out of the bank, Kevin said that the time between 6:00 pm and 7:00 pm was always very good for fishing in the neighboring marshes of the Rio Juruena. I noticed the calmness of the Juruena as the sunset was so graciously reflected from its waters surrounded in the distance by the thick wall of trees and vegetation. In ten minutes we were at the fishing site selected by Kevin. We got out the worms, put them on the hooks and flung our hooks into the Juruena. In less than ten minutes each one of us had a healthy catch. I was very surprised at the quick catches we made in the less than thirty minutes we had been out fishing.

Except for mosquitoes, there were no incidences of piranhas although we had been warned by Kevin to make sure that blood from the mosquito bites did not drop into the river for fear of stirring up the piranhas. As we rowed back to the base camp, Kevin told us that fishing was one of the biggest benefits of living on the banks of the Juruena. It would appear that a river such as the Juruena, as with many of its tributaries for the great Amazon River, had been hardly disturbed through fishing, logging, mining or other man made problems back in 1981. Thus the food chain for fish was still very much intact. This led to an abundance of freshwater fish. It reminded me of the days when I used to go to my grandfather's fish ponds to catch fish with my bare hands in the Niger Delta as the flood waters receded following the rainy season. Personally, I have always enjoyed fishing since my childhood when I used to go to the local fishpond with my late grandfather and uncles. So this experience was very exciting and refreshing in the middle of an intense germplasm collection exercise. While we were emptying our fresh fish into the pans and plates, George strolled down with smiles on his face. He escorted us back to his living room. Shortly after, we went to take our showers and returned for our dinner. Unlike the previous night we went to bed immediately after dinner.

We were up by 5 a.m. as had been the case during the exercise. We loaded up our engine boat. I noticed more gallons of gasoline brought into the boat and the food supply was increased compared to two days earlier. I was fully convinced that this day was going to be a very long one. This was confirmed when our team leader indicated that the site selected for our collection on this day was over five hours of boat ride and that there were some very good prospects. So we went on the Juruena for five hours upstream to the northwest of Conomali. On this river trail, you saw nothing but the waters of the Juruena, the thick wall of forests on both banks of the river and the high blue or cloudy sky above. The place we anchored was called Jair Cardoso. We pulled into a sandy embankment with tree branches hanging over the bank of the river. Using machete our tour guide and tree climbers cut through the thickets. We made our way into the forest onto a footpath used by the local native rubber tappers (seringuiros). We followed the footpath until we came to a very large rubber tree. There was no doubt that this was a very high yielding tree. There were up to 15 cups, about 4.5 liters, hanging on different parts of the tree. We immediately gathered around the tree and took pictures, and marked out a wide area to collect seeds of this very high yielding tree. After spending one hour around this tree we got back on the footpath and continued to the next set of trees.

By the time we completed out prospection for this day, the team had walked more than 25 kilometers on foot. It was already dark as we entered an open area with two huts. The lanterns had been turned on. The rubber tapper introduced us to his young family and after we exchanged greetings our team set out to construct a camping area with all the gears we had brought with us. We shared a wonderful meal with our new host. Like many parts of the developing world this was a family living in the remotest parts of Brazil and barely earned a meager living tapping rubber. So they used the natural things around them as part of their subsistence. Yet they seemed very happy. At about 9:00 pm that evening we all went to sleep. This was one of those 15 hour days that we spent during the four months of germplasm collection.

Next day, we were up again at 5:00 am. We took our camping gears down, had a quick breakfast and started-off on the footpath to the first rubber tree for the day. Seringueiros like other rubber tappers in the world have to start tapping by 6:00 am and be done tapping by 10:00 am. Thus, we accompanied the seringueiro from 6:00 am till he completed all his trees for the day. We collected more seeds and headed for our boat following the same footpath. It was 3:00 pm by the time we set out from Jair Cardoso for Conomali, arriving at about 8 pm. Dinner was light and we went straight to bed, which now seemed very comfortable after a night on the hammock inside the jungle.

Waking up at 5:00 am had become routine and as we entered the sixth day of germplasm collection on the Rio Juruena, our wake-up call no longer came as a surprise. We were always ready to wake up and go. Our trip on this day was another seven hours of boat ride to the collection site. By the time we disembarked for seed collection, we had arrived at Corrego, southeast of Conomali. This was a very long Estrada (footpath or trail). Here, the seringueiro was already at the riverbank waiting for us. He welcomed us with a very broad smile. Then he beckoned us up to a dry area in the thick forest. Again we loaded up with all our camping gears and our backpacks containing an average of 20 kilograms. With machetes the seringueiro and our tour guide cut through the thickets until we came to the clear footpath. All the while, the seringuero kept telling us about the history of the rubber estrada. How old it was and so many rubber trees located in this estrada. With my focus on the rubber trees and the surrounding vegetation I was not aware of what transpired on the forest floor. Therefore, only the quick action of one of our tree climber's prevented me from stepping on one of those orange looking snakes. As my right step was coming down he pushed me forward. I was about to ask him why he pushed me so hard when he pointed to the snake coiled next to the footpath. Having grown-up in the equatorial rain forest of West Africa every snake for me is considered dangerous.

Therefore, once I recovered from my initial shock, I thanked the tree climber immensely for saving me from a possible snakebite even though we had jungle boots on. The tree climber struck the snake on the head with his machete as we apt to do to any snake in the tropics. This incident in no way affected my continued performance on the expedition as it was a normal thing to see snakes in the tropical rainforest and continue with one's life. By 7:00 pm when we set up camp, we had walked close to 30 kilometers on the footpath. As in Jair Cordoso we had our hammocks up and had a light dinner and slept for the night deep in the Amazon jungle. By 5:00 am of the seventh day we were up and ready to go considering the fact that we had a seven-hour boat ride upstream back to Conomali. We continued to stop and pick-up seeds and clonal materials (budwoods).

As the sun was setting on this day, we finally pulled into Conomali. We off-loaded all gears from the boat and packed everything into our little huts. This indicated to me that our river trips on this part of the expedition were basically completed. This became obvious to me when at dinner; George told us the Aero Mateur would delay their flight due to mechanical problems. They would pick us upon the second day rather than the next day as originally scheduled. However, "there are a few rubber trees within one to two hours walk from here that will keep all of you occupied tomorrow," George remarked, with a big grin. Our team leader gave him a look that one could interpret to mean after eight days in the jungle this man thinks we were going to live with him. For some strange reason I had started to enjoy myself and liked the tropical rainforest environment, so much so that I waived at our team leader to arrange for the trip the next day. In fact these eight days strengthened my desire to pursue a Ph.D. in Forest Genetics and Tree Breeding. I felt very much at home and a sense of making a difference came over me.

As we retired to our sleeping areas, our team leader told us that we would make the collections for the next day as long as it would take to complete the collection. During the breakfast the next morning, Kevin whispered into my ears that he wanted to speak with me immediately after our breakfast. Thus, I met with him as the rest of the team put our backpacks together; Kevin walked me towards the area where were we talked about cocoa seedlings. To my amazement Kevin told me that the labor would be available that afternoon to transplant seedlings to a prepared field. With my mouth wide open I shook my head in an excited positive response. I told him that I would be delighted to show the workers how to plant out a plantation of cocoa trees. What spacing was needed and how to handle seedlings that were so old in plastic bags. Suddenly, I realized that this could be a legacy for my team and which if successful could be referenced in the future as an IRRDB cocoa plantation. Then I joked with Kevin that his chocolate farm would be established on that day.

Our trek to Cotriguacu was more successful than we had imagined. We walked for one hour and started to collect seeds. Although these were not very big trees, seed production was very high. I wondered why George did not recommend this area for our first initial visit. Could it be that George had some personal interest for the land in this little village not so far away from him? And why the last minute for us to prospect for the rubber germplasm only after the aerotaxi did not arrive on our scheduled day of departure from Conomali? Be that as it may, our short trip to Contriguacu was very successful with regards to the number of seeds collected. After two hours we trekked back to Conomali, arriving at about 1:00 pm that afternoon.

Approaching Conomali and just outside the tree lines, I observed an open area had been prepared with the traditional slash and burn methods of cleaning bushes in the tropical rainforests. I saw Kevin standing with some of the laborers as they dug individual holes that could take the polythene bags containing the cocoa seedlings. So I immediately joined him. We shook hands, he smiled and said "We have started implementing the plans you gave me to transplant those seedlings" as he pointed towards the cacao plants in the polythene bags. With the excitement of leaving behind a legacy of cocoa plantation in the very remote part of Brazil, I put down my backpack with all its contents. Instead of feeling very tired, I felt a new surge of strength as the adrenaline, triggered by new excitement flowed through my muscles. I took Kevin by his hand, picked up one of the hoes, got one of the cacao plants, and demonstrated the "ball-of-earth" method of transplanting nursery plants of many tree crops. With that demonstration, Kevin provided leadership for the rest of the laborers and we worked until 5:00 pm when I was so exhausted that I told Kevin that I needed to go to our room and clean-up and rest before dinner. He shook his head and waved me off. I stumbled to my room, washed with my usual one bucket of water and then slumped onto my wooden bed and slept until 8:00 pm when I was woken up for dinner.

As we walked towards George's compound where we always had dinner, I asked Kevin why George changed his mind about transplanting those cacao plants. Kevin smiled and said, "You don't know how rejuvenated George has been since your team arrived here. He has never had so many outsiders come to this middle-of-no where to spend time with him. Additionally, you are all interested in doing work with forest trees." Then he paused. And pointing towards me, he said "You have made a lasting impression on George. Since that night you asked intriguing questions regarding his background, he has become very fond of you. Moreover, he never thought he could find an African or someone of African decent or a Negro as he always called them, who was a scientist. In his mind, he thinks you must be one of the best rubber breeders or engineer d'agronome in your home country for you to be selected to come all the way to Brazil. So, when I talked to him about our discussion of setting up a cocoa plantation with those cocoa seedlings he was very excited and told me to get the land prepared as you had told me. I was personally surprised, given his attitude towards those plants in the past. As you can see, there was no problem after that. Thus, with only about a quarter-acre left for planting, George will actually get a cocoa plantation right here in the Brazilian jungle." I felt overwhelmed by these comments. As far as I was concerned, I was only trying to put my knowledge and experience to use.

As we pulled our chairs and sat down to dinner, George announced that he had received radio confirmation for the arrival of our twin-engine plane the next day. The aerotaxi company had found a plane and a pilot. We were going to depart Conomali for Cuiaba about 11:00 am the following morning. Then he pulled out two bottles of wine, placed them on the table and stroking his long grayed beard, he smiled and said, "Tonight we shall have dinner with wine. You have all worked very hard and brought excitement to my little township." Now is the time to eat, drink and be merry. Your team leader told me earlier this evening how successful you have been." By now each of us had been given a wine glass and wine poured into each glass. George got up again and said, "Let us toast Brazil, the IRRDB and all of us gathered around this table." We toasted and then sat down to the most delicious meal that had been served in all the nine days we had been in Conomali. There was so much food and so varied, from fresh fish to "bush" meat and fresh vegetables that I personally did not know where to start. The sight of the food on the table was enough to get filled. However, I picked up my cutleries and filled my plate. I ate and drank to my fill. Since my physiology never went well with alcohol, a glass of wine was all I could handle. It was indeed a very nice farewell dinner.

All the time dinner was in progress I observed that George kept looking at Kevin and I. At one point I noticed Kevin nodding his head as if to answer a question from George. Thus, as soon as dinner was over, George held up his glass of wine and said "Let us thank our young IRRDB friend who has helped us to establish a cocoa plantation here in Conomali. If successful, we shall have a source to make "Conomali chocolate". Thank you young man and thanks to your sponsors. We wish you many more successes and all of you the very best during the rest of your expeditions here in the Amazon Basin. I am glad to note that you chose Conomali to start this exercise. It has been a wonderful nine days for me and my cohorts here in Conomali. Have a safe trip back to Cuiaba and eventually back to your home towns and countries." Then, our team leader stood up and gave George and the citizens of Conomali a big thank you and congratulated him on the arrangements that made our expedition on the Rio Juruena and its surrounding banks and rainforests very successful. He concluded by saying that, members of his team will remember for a long time, if not their entire life, the unique experiences they had gained in this wonderful and not easily accessible part of the Amazon Rainforest. We clapped our hands and stood up to accord a great respect to George for having been such a wonderful host.

As we began to file out of the room, George stood up and gave each one of us a handshake. With me he held onto my hands for a long time, patted me on the back and said, "Son I wish you all the very best. If you are able, we would like for you to visit us again in the future. We would like for you to have a taste of our 'Conomali chocolate'." I thanked him and told him that God willing, I would very much like to return to Conomali in the future. It was a wonderful experience for me as a young scientist. As tears started to stream down from our eyes we waved goodbyes and I walked out of the building to meet Kevin who had been waiting for me. For the first time in many days we walked to our section and room without saying a word. There was a calming effect of the realization that we had become very close friends and it was often difficult for close friends to say goodbye. Or it may just be that there was nothing to say after such a moving and satisfying dinner.

# CHAPTER FOUR

## FIRST TRIP BACK TO CUIABA

I slept very well that night and by the time we woke up the next morning I noticed that our gear was being assembled and radio messages being sent out. We had a light breakfast. Of course, after the type of dinner we had the previous night, it was only natural that the breakfast be light. At about 9:30 am with all my gears packed and ready to go, I told Kevin I wanted to see the Juruena before leaving for Cuiaba. We strolled down to the banks of the Juruena. I stood in front of the engine boat we had used for all the upstream and downstream river trips. I looked up and down the river with its bluish-brown water as it flowed. Just when I bent down to take the water into my hands we heard the sound of what looked like a plane coming from the south. With the trained ear of an experienced air traffic controller, Kevin exclaimed, "That is your plane!! Get ready and lets go to the runway!" Quickly, I reached for the river with both hands, collected the water of the Juruena and threw it up, and watched the wind scatter the drops.

Then Kevin and I turned around and ran to collect my gears. By this time the entire team had assembled in the center of the compound and all the residents came to bid us farewell: the cooks, the gardeners and all the laborers, everyone except George. Then I asked Kevin why George had decided not to be present. Kevin responded by saying that George had been too emotional the previous night and so he woke up early this morning and went out to his little hut in Cotriguacu as he always did when he sought solitude. He could not stand the sight of us actually leaving. In just ten days he had come to see us as part of his kingdom as he now regarded Conomali. As he spoke, I started noticing Kevin's eyes beginning to blink very frequently and he too was looking more and more away from me. Just then, the plane was sighted as it came over the trees and soon touched-down on the grassy runway. It pulled up slowly towards us and came to a complete stop. The engine stopped, the pilot could be seen opening the door of the plane, climbing onto the wings of the plane, then onto the runway and walked towards our group with a large grin.

Our team leader welcomed him and introduced all of us to him. He shook the hand of each one of us and said we had excellent flying weather and should be in Cuiaba within three hours. So the people that escorted us to the runway loaded all our gear into the plane. Meanwhile, Kevin and I stood side by side speechless. Then I pulled-out one of my IRRDB pens and handed it over to him. We both embraced each other. I shook his hands and ran to the plane, because both the pilot and my team leader had started to shout and beckon me to get onto the plane. Also, I realized that Kevin and I were beginning to become very emotional; we could start shedding tears if I stayed to tell him how much I had enjoyed his company. So I ran as fast as I could and climbed into the plane. As the plane taxied on the runway, finally took-off and flew over the trees, I continued to stare below in search of Kevin. There he was still waving. With the carpet of trees below us and the Rio Juruena disappearing behind us, I sat back with my seat belt secured and reflected on my first nine days deep in the Amazon jungle. I had already spent a total of 24 days in Brazil but the past nine days had made the most impact in my life. I had become so attached to Kevin, Rio Juruena and the Conomali environment to the point that I forgot that I was on a scientific expedition to Brazil. As the thoughts went through my mind, I fell asleep.

I woke up to a group of frightened colleagues and pilot. I was sitting next to my Chinese colleague, Dr. Chen. He was older than everyone on the team. I observed fear on his face and eyes. Dr. Chen was about 50 years (at least that is what he told us during one of those treks we had in Conomali). It also explained why he used to slow down considerably during many of the walks through the heavy thickets. But he did very well for a man of his age. My impression was that the average age of the six-member international team was about 33 years. Personally, at the time I was 29 years. I heard Dr. Chen suddenly say "I could have remained in South China and died there rather than perishing here in Brazil." Then looking out the windows of the plane, I observed that there were very dark clouds all around us and that the plane was flying so low that there was not much distance between us and the trees below. Before I could ask questions as to what was happening, the pilot said he would land as soon as he found a landing strip. Furthermore, he said "We are four minutes of flight time away from Cuiaba but the airport has been closed due to the tropical rain storm." I was now wide-awake so I said to him "But you told us in Conomali that we had excellent flying weather. What happened? Were the weather forecasters wrong?" So he explained that in Brazil, the tropical storm clouds often form rapidly and produce storms that were often difficult to predict, particularly since most of these storms often occur in the early afternoon times.

We circled the area for another five minutes which felt like an eternity as the rain and lightning started to intensify. The plane started to bump so much that it felt like a flying kite. I felt we could crash any moment. Therefore, like Dr. Chen, I started to wonder about dying in the Amazon jungle. All the excitement and wonderful experiences of Rio Juruena had given way to a deep sense of bewilderment and concern. Here we were in the middle of no where in a small twin engine plane and all one could see through the window were very dark clouds, approaching heavy rains, mountains on all sides and a carpet of trees just below us. Then I said to myself "I am from the Niger Delta. After all, we have a thick equatorial rain forest there. I would rather die there than to be here where no one knows my family or me. If we crash, the panthers and vultures could eat us up before anyone found our bodies. We could simply become a foot-note in a deadly plane crash rather than this group of courageous scientists trying to advance the natural rubber industry for the world." My thoughts went to my wife and two young boys. Who would take care of them? I thought about my mother who lost her husband (my father) to a ghastly car accident only four years earlier after over 33 years of marriage. It would be very devastating for her to lose her senior son in another deadly accident from the air. What will my friends and foes at RRIN say about the crash given the fatalistic nature of certain traditions of Nigeria. These thoughts grew stronger with every additional bump that the plane went through. Then as a former altar boy, I told myself to think about the church and pray for God's help. God always comes to the rescue of his children. So, I prayed silently as the pilot continued to look for a landing site. Suddenly, with an excitement in his voice, the pilot announced that there might be a landing strip about seven minutes from our existing location, flying north away from Cuiaba where the thunderstorm and lightning were most severe. According to him we could land and stay on this farm until the Cuiaba airport was reopened.

You could see the sigh of relief that showered over our faces. Distance and flight time meant nothing to us at this point. All we wanted was to land and be on the ground rain or no rain. We cheered even though still frightened. Within seven minutes our twin-engine plane touched down amidst a heavy down pour of the tropical rain. The owner of the farm (ranch) brought his old Ford truck and ferried us two-at-time to his house where we waited for the rain to stop and the Cuiaba airport to reopen. The tropical storm ended as fast as it started. We waited another forty-five minutes before our pilot got clearance to fly into Cuiaba. As the plane touched down, we clapped and cheered loudly. We were welcome by officials of SUDHEVEA who had remained at the airport to receive us. We were driven to our hotel, which was one of the best hotels in Cuiaba. We checked-in and I shared a room with our team leader. Flying back from Conomali was very exhausting and three hours ended up lasting seven to eight hours. Hence, as soon as dinner was over we returned to our rooms and slept off. Ironically, after ten days in the jungle, the amenities of modern living no longer meant so much. I turned on the TV in our room to watch the news. Surprisingly, the story of the IRRDB scientists was being reported on this special program carried by the local Cuiba TV. However, staying in an area where you can make contact with the rest of the world was welcoming and refreshing. Yet, these could not replace the very remarkable and rewarding experiences we had along the Rio Juruena.

# CHAPTER FIVE

## INSIDE THE JUNGLE II ITAUBA-ARACATUBA: THE TROPICAL RAIN FOREST-SAVANNAH TRANSITION ZONE

We came into Itauba and Aracatuba as our second and third secondary base camps respectively. Our flight was once again on a twin engine plane organized by our SUDHEVEA partners. Combined, Itauba-Aracatuba area is in the rain forest-savannah transition zone in what is often referred to as the "Cerrados" (more on this later) of Brazil. Itauba is located in the northern part of Mato Grosso. When we arrived at Itauba there was an individual at the airstrip waiting to receive us. This rugged looking individual drove what looked like a 1960s Ford pick-up truck that showed signs of good use and age on it. We loaded our baggages and jumped onto the open bed of the truck. Paulo our team leader sat in front of the truck with Hosea, the name of our new host. As we drove slowly through the jungle with its over hanging lianas and large mahogany trees, we could see large cleared out portions of land containing planted pastures with cattle ranches. The gentleman, Hosea, took us to this wooden made hut prepared for us specifically. This location in Itauba called Del Pai became our base camp *per se* just as Conomali on Rio Juruena. Six seringals had been identified for germplasm collection in this Itauba area: Sr. Lupecio, Sr, Porfiro, Sr. Nome, Sr. Mario, Sr. Benedito and Sr. Sede. According to our new host, these were scouted after they received a radio message from SUDHEVEA about our expedition. Much of the collection was done on foot resulting in us covering an average of 25 to 30 kilometers (15 to 20 miles) daily depending on the length and size of the estrada in each seringal.

On our first day of collection, we woke up very

early at 5:00 a.m. and went to the little stream about half a mile from our wooden hut, took our showers (I often wondered why we had to shower in the mornings considering that we would shower at night after trekking through a very humid and thick understory of the world's largest jungle), and returned to get ready for our trek into the forest for collections. Then we were directed to another wooden hut but much larger than our living area. By 7:00 a.m. we were well underway to the first seringal on the list for rubber seed and budwood collection. We were extremely tired when we got back. After tiredly walking to the stream, our 'newly minted shower" we got ready for dinner which was served at the same place we had breakfast. Walking back to our designated hut I noticed that connected to our dinner place was another large room operating what looked like a corner store.

For the next seven days as we went to the jungle and back to the dinning area set up by the ranchers it soon became clear to me who owned the store and how the workers were being treated by the managers of the ranches. I noticed that the workers mostly native Indian tribes and blacks worked in the various cattle ranches. One evening as Hosea sat around us for dinner I called him to seat close to me and let him understand that I wanted to know more about the ranches containing so many cattle and the tremendous amount of deforestation. Hosea could not speak about those things other than talk about how the corner store operated. "This corner store is the only place where we as workers can purchase modern items of toiletry, soft drinks, cigarettes of all types, clothing and others. This store belongs to absentee landlords who live in Rio de Janeiro or Brasilia. The store was regularly stocked from their larger stores in the big cities. The workers here have to set up an account with the store manager. Every time he purchases an item for his family the store manager deducts the cost of the item from his anticipated monthly wages. Thus at the end of the month, when the worker goes to pick up his pay, which was always done in cash in this far reaches of the Amazon basin, he finds out that he would receive only about five to ten percent of the wage on paper. In extreme cases the worker could even find himself owing money to the rancher." Hosea concluded his statements with tears filling up in his eyes and I saw him looking around as if to check on who may be around listening. I thanked him immensely for that information. I also saw my Portuguese translator look away as we concluded the discussion. The whole set up was quite new and strange to me having never been exposed to something like that in my part of Africa. I couldn't say a word, shocked to hear that some one could work for a month and instead of getting paid will be told "you owe the employer" as you tried to feed and clothe your family during the month.

Over the next several evenings as we retired into our living area I engaged both our Portuguese translator, Jaques, himself a Belgian native, who lived in the state of Bahia, and our team leader, Paulo in debates covering the nature of the Brazilian Government policy and these developments in the virgin forests and its peoples. As we came across a few more Blacks and Indians most of whom worked in these ranches I had to ask Jaques how much he knew about the slave trade and what anthropological information exist to cover these people. Jaques told me about the Brazilian Northeastern State of Alagoas whose population of blacks also parallels those of Para State. In this state, Alagoas, the Quilombos still maintain strong African cultures and reminds everyone of the interactions between the Indians and African slaves. For example, the communities used African-style matchetes and instruments in Indian mixed farming of manioc, corn, rice, oil palm, breadfruit and others. According to Jaques, 'the transatlantic slave trade was a very massive undertaking. Not many people know that for every African who landed in British North America, twelve Africans landed in Brazil to work in gold mines and sugar plantations. It is said that a third to half of these Africans died in these mines and plantations. Over time, communities like the Quilimbos evolved especially after Brazil abolished slavery in 1888. Unfortunately, Brazil was the last nation in the western hemisphere to abolish slavery. Such abolition did not end discrimination, as was the case in the United States and other parts of the world. Quilombo communities came to be regarded as the maroon people of Brazil because of the tendency to move as far out of sight as possible and conceal themselves. These people sometimes called Quilombolas have now emerged in recent times as significant landowners in remote parts of Alagoas. It should be noted that like in Para and other areas of Brazil, well out of the sight of the slave owners was the tremendous interplay of cultures among the Africans and Indians." Jacques concluded. I listened so attentively that because of a personal interest for stories surrounding the slave trade and could have let him continue but the realization that we had to be up by 5:00 a.m. we decided that he could continue the story next evening. Even then, I still put in one more question. "So what is the Brazilian Government doing for these people now?" I asked. Jaques responded this time in French. "Le gouvernement brésilien est élaborer des politiques pour déclarer des endroits comme Palmares dans l'Etat d'Alagoas comme un parc National, une zone qui compterait près de 20 000 à 30 000 africains et Indiens. Il a un développement similaire pour d'autres Etats, en particulier avec la prise de conscience internationale du bassin amazonien. En fait, c'est à l'honneur du gouvernement brésilien qui est maintenant prévu de mieux intégrer la population noire du Brésil dans l'économie brésilienne plus efficacement. Au cours des 20 à 30 prochaines années, cette intégration devrait progresser considérablemen." "Merci beaucoup, Jaques" I responded with excitement. For me it was "Living History" which was never covered in history books in secondary schools or in the mandatory history courses we took at the university for my undergraduate degree. We have heard about the great

Brazilian football teams and great players like Pele, but no one has ever given me this type of memorable Brazilian history.

It is 6:30 a.m. and we were on our way again to the next seringal the next morning prepared for another 20 to 30 kilometer trek when I walked up to Paulo and told him that there was very little genetic variation among the seeds and trees we had seen so far. Paulo assured me this particular seringal to which we were going had some interesting trees but that we must arrive there while the tappers were tapping their trees. So I walked along the estrada with great anticipation. We stopped at the first tree just about ninety minutes after departing our base camp. Like most of the trees we had seen in the other seringals the seeds were not too different. However the tree was a little larger. Three hours walk from this tree brought us to the anticipated seringal. As we walked around the seringal we arrived at a group of rubber trees with many seringueros still collecting latex from the previous days of tapping. Paulo looked at me and smiled. It was as if he had found an answer to my question earlier in the day. Our two tree climbers were already rushing to the center of the seringal. About a hundred feet I could see this rubber tree towering above every tree around it. Everyone was streaming around it. Upon arriving at the tree, I could immediately see why our team leader was smiling with such confidence. I looked at him now and shook my head and nodded with great appreciation. This was a very high yielding tree. We counted the cups hanging from almost every part of the tapable portion of the tree. The trunk was average for a rubber tree, about 12 to 15 years old the prime productive years of a rubber tree in a plantation. I counted the number of latex cups, which came to 15. Although these cups were made out of coconut husks, yet they seem to contain the same equivalent volume used in regular rubber plantation cups. We clustered around the tree and positioned ourselves for pictures even before we started taking notes and collecting seeds and budwood materials. After taking pictures, our tree climbers got themselves ready working with the seringueros to climb to the very top of the tree. We wanted as many budwoods as we could get from this unique tree in the Itauba collection site a relatively dry area compared to other places where we had been up to this point. I was afraid that the tree could face the possibility of "slaughter tapping" in the future. Meanwhile, Paulo, Jaques, Dr. Chin (our Chinese team member) and I collected any and all seeds within 80 to 100 feet radius of this tree. The excitement level was so high that we spent nearly three hours in this seringal. It was already 4:00 p.m. and because of a few more trees to visit we left all our collected materials at the site of this particular tree. This was a very rich colocacao for these wild rubber trees. The sun was already setting when we started making our return trip to Del Pai, our base camp. This meant we were going to work nearly three to four hours in darkness back to the camp through a mixture of jungle and high grass patches until we came to the ranch. No doubt, it will be an understatement to say that I wasn't feeling fearful as we walked back to the camp in that darkness with the moonlight piercing occasionally through the "light gaps" overhead in the humid forest.

Everyone, including Hosea and all those in this make-shift village were out with their lanterns waiting for us as we emerged from the dark forest into the open pastures of the of the ranch. It was obvious that they were extremely worried that here was a group of scientists who knew nothing about how these jungles operated dared to be working very late into the night. There were natural dangers of poisonous snakes and animals of the night and the various intertribal native Indian fights, but more recently the undeclared wars among ranchers scrambling for fertile lands for pastures for their cattle and pitting bloody showdowns among powerful absentee elitists landlords who are clearing forests for agribusiness against family farmers clearing small patches for crops but are still dependent on the natural forests around them for sustenance and survival. Hosea could not hide his feelings as he walked up to me and asked me why we had spent so much time in the jungle and not returned to the camp before sunset. Exhausted and as we slowly walked towards our hut I explained to Hosea that this day represented one of those days why the IRRDB and Brazilian government were spending so much money to have us here in Itauba. Furthermore, it was the type of day when a scientist finds the genotype of a plant species he is working on. As I spoke, new adrenaline started flowing through my veins and I told Hosea that it was the excitement of seeing the many rubber trees in the wild which were not only highly productive but also showed much variation compared to what we had seen in the previous three days. I entered into my section of the hut as Hosea and others reminded us that a late dinner had been served directly into our hut so that we could eat and go to bed immediately in our hammocks. Although tired and sleepy, Paulo held a very short meeting in which he indicated that because of the large amount of materials brought in from Benedito and Mario seringals we would not go out the following day but spend the entire time preparing all the materials for dispatch to CNPD in Manaus as we had always done at the end of each day of collection.

We woke up late and it was already sunny
and bright in the Amazonian non-rainy day. I quickly went down to the stream and completed my morning toiletry including showering. Breakfast was served and the team was assembled in the dinning area for nearly sixty minutes to reflect on the previous day's excited activities, discuss our work for the day and plan for the remaining two seringals in the Del Pai base camp. Then  with a sarcastic smile Paulo turned to me and said "Kome, are you still looking for genetic variation in Itauba?" I smiled and replied sacarstically as well "Yes, of course. Isn't that why we are out here as scientists?" With breakfast completed and a little relaxed we went back to our hut. Our tree climbers searched for a tree with a nice shade where we could seat,

make notes, sort out the seeds and prepare the budwood prior to being dispatched. We worked feverishly to make sure we had the items ready because the plane that came to the ranch to deliver supplies and take materials back to Cuiaba was expected to arrive by 3:00 p.m. It was this plane that was to take our materials as well. We worked through lunch with snacks and soft drinks being brought to us from the corner store. Sweating under the heat of the humid shade and heat the seeds were sorted for broken and diseased ones, budwoods were sized and all items were placed in Styrofoam boxes after labeling. We all walked to the landing strip with our tree climbers carrying the Styrofoam boxes. Although an hour late we were very pleased to send four full days of germplasm collection. As we backed to our hut, we were reminded that the next day was a Sunday and there will be only half day of work especially as the remaining two seringals and colocacao did not have as much variation as the Benedito and Mario.

That evening, as has been the case since Conomali, Paulo called us to a strategic planning meeting on the basis of the information he had just received. In addition to the two seringals Sede and Nome which we were going to work with prior to moving across the "Cerado" area between Itauba and Aracatuba, we were supposed to visit Utiariti in the Rio Tartaruga region which is not too far from Aracatuba. But a last minute radio message had been received at the ranch asking us to stay away from the area even though it had a lot of seringals with as many as 500 to 1000 wild rubber trees in various clusters. He concluded by saying "by the way, our transportation Monday will be by a 1972 Massey Fergurson tractor." We all looked at each other and wandered where the seats were located on the tractor and why we were going to ride a tractor. Whatever happened to the Ford pick-up truck that brought us from the Itauba airstrip five days ago? Regardless, we were in the jungle and nothing we can do about it. Paulo reminded us that this was part of the challenge of germplasm collection. Our meeting was over and we went to dinner in a more relaxed manner.

I waited for Paulo to be alone just prior to us going to bed. "Paulo, I was very interested in the comment you made with regards to us not being able to got to Utiariti. What is the problem? I think we could go there and then look for a boat that will take us on the Rio Tartaruga to Aracatuba once we finish collection on Monday. I do not know the exact distances but according to your presentation we are not too far and that in fact Utiariti is actually equidistant to Itauba and Aracatuba. Furthermore, I read some excitement in your voice when you briefly mentioned the potential genetic variation available. There has to be a more critical reason for us not going there." I concluded my statements and questions and waited for Paulo's response. "Kome, I just don't know why you ask all these questions. You are a Plant Breeder. Why don't you simply stay with questions regarding the collection exercise" Paulo remarked. "Yes, I am a Plant Breeder. But as you can see everything we have done so far has been based on the cooperation of people whom we don't know and who SUDHEVEA had scouted to make our expedition successful. I am also having a great life experience. I have loved being here in Brazil. So please don't be upset with me." I responded.

"Okay, Kome" Paulo said while adjusting his sitting position. "You see, in Brazil there are groups of Indians deep in various jungles of the Amazon Basin where the government has prevented explorers and all foreign visitors from reaching. Indian tribes like the Kanamari, Flecheiros, and other remaining pockets of isolated Indians. It has been stated that external accessibility to Indian tribes for nearly 500 years, has resulted in the Brazilian Indian population being reduced from over one million to just over 300,000 because of disease and other devastating events." Continuing, Paulo said "Therefore, despite the large number of seringals in the area, the Utiariti Indians have not been exposed to the rest of the world and hence an avoidance of diseases and pests. SUDHEVEA is not about to let us be the first to do so. Many scholars have indicated that like other native Indians, isolation has become a protective and defensive mechanism for the Indians. It offers these groups the best hope of preserving cultural vibrancy and survival. They have become the tenacious protectors of the pristine forest. It ensures ecological preservation of millions of acres of biodiversity in the Amazon Basin." Paulo ended his answer with a look that would show that "I hope this satisfies Kome." In my mind, I was tremendously pleased with his response. Beyond being a Natural Rubber Breeder I was also very interested in the preservation of genetic resources especially of hard wood tree species such as mahoganies. This became a personal thing for me following the demise of the African Timber and Plywood Company (AT&P) in Sapele, Delta State of Nigeria. I have always wondered, how do we genetically restore lost species that take 80 to 100 years to fully establish themselves with the kind of wood quality they provide and the associated plant and animal species that grow around these tree species. So I turned to Paulo and thanked him immensely for his explanation "Paulo, I am extremely grateful for your thoughtfulness and dedication to the scientific community. You are a true scientist given your patience." I concluded my statement and went to my hammock for the night.

We undertook our collections in the two seringals that Sunday morning with very little excitement. These two seringals appear very much like our earlier two days. The rest of that Sunday afternoon, I took a walk around the ranch to see how people lived and how they got along. Also I went to some of the paddocks and probed how these paddocks were being created. At one spot I saw slash and burn in full operation. Fire had been set in one part of the forest while in another part there were close to 20 laborers using machete to cut down midsize trees and clearing the underground growth exposing the soil to further degradation. Although pastures were being established, this was unlikely to keep up with erosional issues in an area so close to a part of the Brazilian Cerrado. At dinnertime I was back to my hut. I joined the rest of my team for dinner. Paulo asked us to be prepared for departure at 8:00 a.m. the next morning.

I watched as the 1960s Massey Ferguson tractor pulled up. A little wooden cart was hitched onto it. Our luggages were put in the cart and our two tree climbers climbed onto the remaining space in the cart. I wondered why the remaining five of us were going to seat. The tractor guided Paulo onto one of the motarboards while Dr. Chin sat on the left side motarboads. Jaques and I were made to seat in a semi-squat position on both sides of the tractor operator. It was a very unique way of traveling in a remote part of Mato Grosso. We drove off on an equivalent speed of about ten miles per hour. We slowly transited from the rainforest on the Itauba side of this region to the Aracatuba side taking about five hours to complete our journey. Thus for five fours we went through the small section of the Cerrados in the Amazon Basin. The Cerrado is a large tropical savanna ecoregion of Brazil. It is the southern hemisphere equivalent of the northern hemisphere wooded savanna in sub-Saharan Africa. It is the second largest habitat after the Amazonia accounting for a little over 21 percent of the Brazilian land mass. It is also considered one of the richest tropical savanna regions with high levels of endemism and a wide variety of animal and plant biodiversity. Climatically, the Cerrado has two dominant seasons, wet and dry. Vertebrate and insect biodiversity have been reported at length for the Amazonian Cerrados. Although soil fertility is high in the undisturbed cerrado, continuous farming can result in rapid depletion of nutrients. Yet this region has been targeted for agribusiness development in Brazil (discussed later in this chapter).

Once again, I asked both Jaques and Paulo what was the potential for agricultural production in the Cerrado. Paulo who is the plant scientist took over the discussion. "For many years, the cerrados were thought to be worthless for Agricultural Production. But preliminary research by EMBRAPA, indicate that this region could represent the area where Brazil could develop and become a world agricultural power in 20 to 30 years. As you know, Brazil has already declared an energy independence policy based on the fact that a number of cars have started to run on alcohol produced from sugar cane. The oil embargo of 1973 and the oil crisis of 1979 following the Iranian revolution made the military government of Brazil to declare a policy of energy independence. This is now the trend. Similarly, the vast arable lands Mato Grosso and Mato Grosso du Sul are areas EMBRAPA believes Brazil could use to become a world Agricultural Power. This has been given credence by Dr. Norman Borlaug, the Nobel Peace Laureate who described the Cerrado as one of the Earth's last remaining arable frontiers for the expansion of agriculture. EMBRAPA is engaging its scientists of various disciplines and specialization from soil scientists to plant breeders to chat a course that will make this a reality." Paulo concluded. "This is very fascinating, Paulo. I will be looking forward to revisit Brazil in the future when it becomes a great World Agricultural Power." I responded.

Our tractor operator pulled into an area with two tent-like structures. We had already passed the cerrado into another tropical rainforest along the Rio Ferro. The sun was setting when we got here. All along the way passing through dirt roads we had seeing loggers and logging trails. This is Aracatuba located just south of Sinop/Itauba. Aracatuba area is on the banks of Rio Ferro, a much smaller river compared to Rio Juruena. It had a lot of lumber processing mills and because of the terrain much of the collection we made was done on foot covering an average of 25 to 30 kilometers a day. We were beginning to off-load our bags when some one ran quickly to stop us. He had a short list in his hands and something that looked like a hand radio. Paulo went up to chat him. Thereafter, he welcomed us and then walked with us to a wooden house just sitting on the outskirts of what looked like a living quarter for some logging firm. The house had a general living area (common room) and two separate bedrooms containing two twin sized beds with grass mattresses in each room and one shared bathroom. It was obvious that this building had been prepared for us as scientists. The tents were reserved for our tree climbers, Jeremias and Silva. It was getting dark, so our new tour guide Luis had to take Jeremias and Silva to the tents. Our tractor driver would use one of the tents and return the next day.

We were up at 6:00 a.m. the next day. Breakfast was served in the common room. At about 7:00 a.m. Luis, who welcomed us the previous night came in with another gentleman much older than he was. In my mind, I thought Luis was in his late 20s while this gentleman may have been in his early 40s. "Good morning, gentlemen," he said in a very deep voice for a five feet six inches tall man. My name is Joanos and I mange the mill here in Aracatuba, which is not too far from here. In fact if stand on the porch you can see the top of the mill. SUDHEVEA contacted us to have you as our guest for up to four days whenever you get here. So, we are very happy to welcome you and appreciate what you are trying to do for the natural rubber industry." Then looking at each one of us as if to be sure he was safe to say it he joked that "We wish those Malaysians did not come to steal our rubber seeds we will be having a lot of natural rubber plantations!!!" We all laughed knowing fully well what the actual story was. However, it represented the perception many Brazilians had even in these remote parts of the Amazon basin. Then pointing to Luis, Joanos said, he is going to be your guide through out your stay here. He will be checking on you from time to time. Make sure you keep him busy because we pay him very well." He smiled shook our hands and left us with Luis. We sat down with Luis and developed a plan to visit three colocacao, Rio Ferro, Corrego

Banderaute and Corrego Tartaruguinha all of which we would cover by foot.

Just as we were settling into our new wooden huts that afternoon after a preliminary walk around Aracatuba, Paulo informed us that we would be halting our expedition temporarily on Tuesday to return to Cuiaba for Carnival (Chapter Six). We had been out for eight days already and so this came as a nice break but I wondered why carnival was so important that we had to halt our scientific work, even though temporary. This not withstanding we prepared for departure even as we took advantage to visit the nearby Estrada, Corrego Bandaute which was less than three miles away from Aracatuba. The Estrada was very long but we were able to cover it just before it got dark. We cameor back to our wooden house to have a late dinner. That night, we had to make a decision as to what to leave behind considering we were going to be returning to complete the remaining colocacao with all the seringals identified.

Luis was waiting for us when we got back from Cuiaba and the carnival. "How was your carnival?" Luis asked. Without hesitation Paulo replied "Very good and I think my African colleague here can attest to that." Luis looked at me and nodded in the affirmative with a big broad smile. I asked Luis if he also went out for carnival. He said yes and enjoyed it very much as it was a nice break from the intense activity here at the mills. Because we had a whole region to complete and two colacacao here in Aracatuba to handle, we were given only two days with a third day to organize collected materials for shipping to Manaus and flying back to Cuiaba. Therefore, as early as 5:00 a.m. the next day we were up, had an early breakfast and by 6:45 a.m. the collection team was on its way. We walked through the small town of wooden houses and stepped into the jungle en route to our first seringal. This Estrada turned out to be very long. It was mid day, we had stopped for early lunch using our parked lunches. We came to one of the creeks draining into Rio Ferro. On this trail the creek was too deep to swim through because the currents were running very fast. And it was just a little too wide to jump across. We walked half a mile on either side of the trail cutting through lianas and thick undergrowth looking for a narrow spot and perhaps shallower which we could wade through as we had done in some creeks around Rio Juruena but we could not find anything better than our trail or Estrada. So we came back to the original spot. The report of excellent wild trees beyond our current location made it extremely worthwhile for us to reach those seringals. One of our tree climbers, Jeremias spotted a tree close to the edge of the gully-like bank of the creek very close to our original trail. He called on his colleague, Silva. They both evaluated the height of the tree and the surrounding under growth. I saw Jeremias walk up to Paulo and said something to him. Paulo nodded in the affirmative. Jeremias carefully cut the tree and guided its fall so that the trunk was completely laid across the width of the creek. They tested its strength to ensure that it would take our weights as each one crossed it. Then they cut another smaller but more flexible tree right besides it to form a hand railing to be held for walking over the bigger trunk. With our emergency bridge "constructed" we were ready to cross. But the question was who would go first. Dr. Chin, our older Chinese colleague volunteered to take the lead. Silva followed so as to wait on the other side in order to provide assistance. Jeremias took the rear, crossing last after everyone was secure on the other side of the creek. When we arrived at the three seringals in this colocacao, the amount of genetic variation observed, made the "bridge construction" worthwhile. We asked the seringueros a lot of questions as to why the trees looked so healthy. These were perhaps the best looking wild rubber trees we had seen up to that point of the expedition. We returned to Aracatuba and the wooden house, elated about the collections made.

On the final day of collection we once again left our living area at 6:30 a.m. We trekked and trekked for nearly ten kilometers before finding a seringal for collection. When we left this seringal, we went through a very thick jungle and came across a very nice stream with clear water winding its through the forest. We could see the clear gravel and sand below the stream. We stopped to admire the cleanness of the stream. Coming from the Niger Delta, I recognized quickly that this was one of those drinkable waters. So I took a squatting position and scooped some of the water in my hands. It was very cold and because I was also very thirsty, I drank to my satisfaction. Then I invited every one to enjoy the gift of God in the middle of the Amazon Rain Forest. Everyone laughed and scooped their own amount of water and drank to their fill. The water was so refreshing, I emptied my water can and filled it with this fresh water which I drank for the rest of our collection on that day.

The estrada we were on led us from inside the thick jungle on to what looked like a logging road that had been constructed by loggers. Standing son the edge of the jungle it wasn't long before we saw one of the logging vehicles loaded with a huge log that looked like a huge mahogany lumber. We stopped the driver who told us that a private company was in the area not too far from where we were felling trees to be taking to the sawmill in Aracatuba. These made me to turn to Paulo and Jaques again with questions about these roads that may be causing deforestation and fragmentation of the forest. Again, Paulo did not hesitate to respond. "Yes, Kome. Although the Brazilian government is in the process of building highways across the Amazon basin, except for a few federal and state highways including the east-west Trans-Amazon Highway, the BR-163, which splits the heart of the Amazon from southern Mato Grosso north to Santarem in Para, almost all roads in the Amazon are unauthorized. In fact both major roads are their initial phases of construction and it will be several years before they are officially completed. Meanwhile it is estimated that there are well over one hundred thousand miles of the illegal roads most constructed by loggers to fell mahogany and other important hardwoods for the lucrative export market." Paulo concluded. I told him that the same thing was happening to mahoganies in West Africa and that often there is much more damage done to the microclimate and ecosystem when just one mahogany tree is felled.

We trekked back to our wooden house in Aracatuba having completed our collections in the Itauba-Aracatuba second and third assigned sights for the Mato Grosso team. As we walked back through the jungle, I told myself that I would have loved to visit the sawmill but because we had to prepare all the collections for shipping and then get ready ourselves to fly back to Cuiaba in preparation for another lengthy expedition in Villa Bella on the Rio Guapore. That evening we ate and went straight to bed. Having asked so many questions during the day, I did not want to border Paulo or Jaques. I could see the strain of two days of none stop trekking on their faces. I was also very tired. Our departure the next day was set for 2:00 p.m. So, immediately after breakfast along with Jeremias and Silva who came from their tents we got all the seeds and budwoods made ready for traveling back with us to Cuiaba and then be flown to Manaus. Luis joined us for lunch. We asked about Joanos. He said Joanos could not make it to see us off because of other commitments but he sent his greetings and wished us all the success in our continued germplasm collection. We asked Luis to extend our deep appreciation to Joanos for the tremendous support they gave us including keeping our items during the carnival period when we were in Cuiaba. Luis check his hand held radio and said our plane was a few miles away and that we could start moving to the airstrip. Our pilot touched down at 2:00 p.m. and within 15 minutes we were in the air. Every time I get on this single engine plane I am reminded of our first trip back to Cuiaba from Conomali. Of course this flight was smooth and we arrived safely and checked into our regular hotel referred to as our primary base camp in Cuiaba.

# CHAPTER SIX

## BRAZILIAN CARNIVALLO: CARNIVAL IN BRAZIL

As we boarded the plane early the next morning I noticed the smiles and excitement on the faces of our Brazilian counterparts. I was clueless as to why our Brazilian colleagues were in such unusual happy mood in the middle of germplasm collection. Furthermore, we were expected to go back to Aracatuba for the completion of the collection exercise in the very unique region of the Rain Forest-Savanna transition zone with its cerrados and specialized ecological biodiversity. Upon landing at the Cuiaba airport, the air was filled with excitement and everyone seems to have been in a festive mood. This mood intensified as we checked into our regular hotel in Cuiaba. Paulo and I have been roommates since arriving in Mato Grosso so I asked him two questions." First why did we have to cut short our collection in Aracatuba, and secondly why is everywhere so decorated as if this was Christmas season?" Then he started to laugh before answering my questions. "Kome, I see you have not heard about the Brazilian festivities related to Easter. It lasts a full week and all Brazilians as a matter of tradition do participate in the CARNAVALLO!!! Over the next three nights, I shall take you to the city center to see the dancing and costumes on display. For Brazilians, it is viewed as a holiday for free expression, in honor of the life, death and resurrection of our Lord Jesus Christ." The Brazilian carnival rejoices life. It spreads euphoria through its dance, song, music, food and people." I looked forward in great anticipation to the first nightly CARNAVALLO.

At about 5:00 pm the next day, Paulo woke me up to get ready for dinner and CARNAVALLO. We had a sumptuous meal considering what our meal had been in the last few days in Aracatuba. As soon as the meal was over, we took a walk to the City Center. As we walked down I saw a lot of people moving in the same direction. There were people with all kinds of dresses, costumes and attires. Suddenly, I started feeling the same level of excitement. By 5:30 pm we were at the City Center and the program activities were already under way. Singing and dancing permeated the air, drums of all kinds were beating, guitars and trumpets blared and by 7:00 pm the entire center in Cuiaba was filled, one could hardly find a place to stand or sit. Then the actual dancing, moving from east to west started. It was extremely colorful. There were different types of dances more than I had seen up to that point. One of the most stunning part was the near nakedness of a number of dress types worn by men and women particularly so for the women. Of course, others wore flowing gowns and dresses that covered all body parts. As the music played and people danced, the crowd went into a frenzy of shouting and singing and also joined the dance in as much of the space available. At one point I was carried away by the music from the drum beat of the carnival as I started moving my head and legs from the Camarotes (ringside seats) we were sitting on. Paulo at this point joked that I should go down to join the dancers. It was indeed a joy to watch the costumes and the flow of human beings like a sea of water. We watched till about 10:00 pm. Exhausted we stumbled back to our hotel. All around us were people dancing back to their homes. As we entered our room, I asked Paulo, how did this kind of festival become part of the Brazilian culture? In response Paulo took me through a lengthy discussion. "Historically, Carnival has been celebrated in Brazil since about the 1600s in Rio de Janeiro when the city's bourgeoisie of mostly Portuguese origin imported the practice of holding balls and masquerades from Paris. Previously, it mimicked the European festivals but gradually absorbed and creolized elements from Native Indian and African cultures.

Carnival literally means to remove meat, a practice undertaken by Christians in particular practicing Catholics, during lent while practicing repentance in honor of the death and resurrection of Christ. The genres of the music for the Brazilian Carnival are the samba-enredo, the samba de blocco, the samba de embalo and the marchinha in Rio de Janeiro representing the Southeast region; the frevo, the maracatu, the samba-reggae, and Axe music in Bahia and Pernambbuco representing the Northeast Region. Carnival styles also vary from state to state. These include the Rio de Janeiro style, the Sao Paulo style, the Bahia style, the Pernambuco style, the Minas Gerais style and others such as Curitiba, Florianopolis, Camboriu and Porto Alegre style."

Paulo paused for about two minutes, took a deep breath and continued." The Bahia style is closer to African cultures so let me describe it for you. Continuing, Paulo said. "According to the history of the state of Bahia "In the 1880s, the black population commemorated the days of Carnival in its own way highly marked by Yoruba characteristics, dancing in the streets and playing instruments. This form was thought to be "primitive" by the "upper-class white elite," and the groups were banned from participating in the official Bahia Carnival, dominated by the local elite. The group defied the ban and continued to do their dances. By the 1970s, four main types of carnival groups developed in Bahia: Afoxes, Trios Electros, "Amerindian" groups, and Blocos Afros. Afoxes use the rhythms of the African inspired religion, Candomble. They also worshiped the gods of Candomble, called Orixas (I thought he was referring the Yoruba goddess called Orisha). An Electric Trio is characterized by a truck equipped with giant speakers and a platform where musicians play songs of local genres such as axe. People follow the trucks singing and dancing. "Amerindian" groups were inspired western movies from the United States. The groups dress up as native Americans and take on native American names. Blocos Afros, or Afro groups, were influenced by the Black Pride Movement, in the United States, independence movements in Africa, and reggae music that denounced racism and oppression. The group inspired a renewed pride in African heritage. Presently, Bahia's carnival consists mostly of Trios Eletricos, but there are still Blocos Afros and Afoxes. The carnival is usually held in Salvador, the capital city of the State of Bahia. It is also called the "Capital of Happiness." According to the Bahian superstar Cariinhos Brown "We play, not for money, but celebrate happiness. Our carnival is a street carnival. It is for everyone, not just for those with money."

"The major music for nearly all carnival is the SAMBA which originated in Bahia from the African rhythms. It reached Rio de Janeiro in the 1920s and it is still one of the most popular styles in Brazil, together with Samba-pagode and Samba reggae (the band Olodum from Salvador da Bahia made samba-reggae famous). From intimate samba-cancoes (samba songs) sung in bars to explosive drum parades performed during carnival, samba always evokes a warm and vibrant mood. Samba developed a distinctive kind of music in the 1930s and eventually became an "official music" of Brazil. Led by Ismael Silva, a group of musicians founded the first Samba School, Deixa Falar in Estaciode Sa neighborhood of Rio de Janeiro. There are now several Samba schools in Brazil. A samba school combines the dancing and party fun of a nightclub with the gathering place of a social club and the community feeling of a volunteer group. At the Rio Carnival, samba schools parade in the Sambodromo."

Paulo concluded his very long but exciting discussion. I listened very attentively reflecting on what we saw earlier in the evening. Then I asked him if Cuiaba had any special carnival like Rio de Janeiro or Salvador. He indicated that the people of Cuiaba observed a variation of Rio de Janeiro and Sao Paulo Carnivals. But because the Bahia style and the sambas have so permeated all of Brazil, there will be elements of the Bahian style throughout all the carnivals. Besides music, there are costumes, food and drinks that completed the entire carnival. The costumes and masks are of extreme importance, as they depict their own story. Their creation requires a high level of understanding of the roles they play. Like music, even the costumes are influenced by the African heritage. Fabrics of different textures, mostly adorned with feathers and sequins, are used. While designing such elaborate costumes, the freedom of movement to the dancers is an essential aspect. Carnival make-up creates an enchanting persona. No carnival is complete without the Brazilian national dish, Feijoada, a pork and bean stew. Streets are full of food cooked either in houses or restaurants. The cuisines differ from place to place with each adding its unique flavor to its carnivals. Dancers eat light meals, but enjoy the full course after the parade is over. Drinks are equally important, to beat the heat. Brazilian soda "Guarana" and beer are the preferred choices.

Paulo also added "Carnival is no longer limited to Brazil. Today, it is widespread in the Caribbean Islands especially in Trinidad and Tobago where I understand that they have a very robust carnival celebration that may rival that of Brazil. Also, I hear that even in the United States of America, the state of Louisiana celebrates Carnival under what they call Mardi Gras in the city of New Orleans. As I said earlier, it has become an annual festival for Brazilians and celebrated from four to six days preceding Ash Wednesday. Its religious flavor marks the beginning of the fasting days of Lent. Furthermore, we Brazilians consider it the "Greatest Show on Earth." It is a huge foreign exchange earner for Brazil. In Rio de Janeiro for example, annual attendance by foreign visitors is estimated at 400, 000 to 500,000 tourists." I reflected for a minute on the description of carnival and realized that similar festivals were held at harvest time in the villages of the Niger Delta. Chief among this, is the "Yam (*Diascora species*) Festival" in which thanks are given to God for a successful harvest season through dancing and the display of different types of costumes in the streets with all types of masquerades especially yam shaped masquerades. There is always happiness in the air during the four days of celebration. These festivals always take place in early October at the beginning of the harvest season. The word "samba" may in fact have originated from Africa.

The next day we enjoyed another bliss at the City Center and continued doing the same for almost three days so that by Saturday night we were completely exhausted. Ironically on Sunday, the actual Palm Sunday we slept late and failed to attend mass. By the time we woke up that Sunday it was already 1:00 pm. Paulo and I had lunch in the hotel and went out for a walk and reflected on the Brazilian Carnivallo. Settling down for dinner that evening, we realized that it was time to go back out into the hinterland for Hevea germplasm collection, our raison d'etre in Mato Grosso.

## *Carnival in Brazil*

**(photo source: article.win.com, 2011; public domain)**

# CHAPTER SEVEN

## INSIDE THE JUNGLE:
## III. VILLA BELLA – RIO GUAPORE

We spent twelve days in Villa Bella along the Rio Guapore on the Brazilian-Bolivian border. Like the Rio Juruena, Rio Guapore also was in the far recesses of the Amazon Basin. This area became our fourth and final secondary base camp in Mato Grosso. We had already completed three quarters of our expedition, so our expectations were the same. Our flight was with another small single engine plane but for much longer distance. It took us about five hours from Cuiaba to arrive in Villa Bella. Upon arrival we were immediately checked into the local hotel. I noticed with interest as we drove from the airstrip to the hotel that this town was in a very remote part of Brazil. The major roads were paved even though there were a lot of potholes on them. Our hotel was located in the city center in a typical Brazilian city with its town-square characteristics. Our hotel was located on the west side of this square overlooking the major road that led straight to the banks of the Rio Guapore. Also striking were the many people of African descent in Villa Bella. In many of the other places we had been to, I did not see that many black people, so this came to me as a bit of a shock (more on that later).

Our team leader wanted us to arrive on Thursday of that particular week so that we could have a day and a half of when people were still at work in order to prepare for our eight-days boat trip on the Rio Guapore. So, the next day we met with the SUDHEVEA local officials who took us on a brief tour of the city, it was not just another sight seeing tour, it was historical. The official told us that Villa Bella was at one time not only one of the largest cities in Brazil, it was in fact the country's capital, when the Portuguese ruled Brazil. This opened my eyes and many of the initial questions I had in my mind about the city were cleared. Now it was obvious why the roads were paved and why the houses and the buildings were the way I had seen them constructed. How could this city so far away in the heartland of Brazil be the capital city? This was very interesting to me. Our tour of the city took us to the only commercial bank, the city hall, the Agriculture Department, and finally to the small Inland Waterways Department located on the banks of the Guapore. As we made our way back to the hotel on foot I noticed that many of the vehicles were very old and unlike the new Ford and Volkswagen vehicles I had seen in Rio de Janeiro, Brasilia and Cuiaba, these Ford vehicles were mostly pick-up trucks that were rusted and had worn-out tires. Yet this abandoned capital of Brazil still held its own.

That Friday evening, one of the SUDHEVEA officials took us out to dinner in the City Center. We were treated to an excellent and very delicious Brazilian meal. I noticed right away that the black people were mostly cooks and servants and that they stared at me sitting in the midst of all these white Brazilians. I was so embarrassed that I told our SUDHEVEA official that I would like to know more about the history of the black people in Villa Bella and in Brazil for that matter. How come in this city with what appeared to have a high population of blacks, there wasn't even one of them as a counter cashier at the bank nor had I seen any one of them in any meaningful position in the various offices we had visited. Then he gave me a fantastic history of blacks in Brazil indicating that the blacks in Villa Bella were ex-slaves of large farm owners back in the early days of colonialism and slave trade. Because the Portuguese could not fully rely on the labor provided by the native Indians, they had to obtain slaves for this purpose. Thus, the black population of Brazil grew very fast and today makes up 30 to 40% of the Brazilian population. Yet many blacks were underclass citizens so when in a remote and poor town like Villa

Bella blacks see another black person mingling with white people who hold all the offices in town, they are not only in awe of the individual but also suspicious at the same time. He concluded by saying that I should not feel worried, but go about the job I had come to do in Brazil. However, deep down I felt that this people were not etting a fair treatment if they made up so much of the population. By now, we had reached our hotel. So I thanked him very much for the information. We bid each other goodnight as he wished me all the best during the rest of my stay in Brazil.

The next day was Saturday. At about 9:00 am following breakfast, our team leader requested my Thailand colleague and I to accompany him to the stores to purchase items for the anticipated eight days trip on the Guapore. I did not hesitate to accompany him as I saw this as an opportunity to see much of the shopping areas of the town. As we moved from store to store I noticed that our team leader was picking-up large quantities of cigarettes, boxes of whisky and other alcoholic drinks in addition to our regular field supplies. I asked him why he was buying all these things that had nothing to do with germplasm collection. Furthermore, after these many weeks I have noticed that other than our Chinese colleague none of us smoked or drank alcohol. He looked at me and smiled, and said, "Wait and you will see what happens to these drinks within the next 48 hours." My Thai colleague and I assisted him to carry the purchased items to the hotel.

Early the next morning we assembled at the wharf where the covered motorboat was being loaded with all our travel gears. As soon as the loading was completed we entered the boat and the boat mechanic cranked the engine and pulled off into the main course of the Guapore away from the banks. A number of Villa Bellans came to wave good-bye. As we entered into the downstream flow of the river, I noticed that there were as many additional people in the boat as there were germplasm collection team members. Besides the mechanic, there was a cook; there was a repairman and an additional tour guide. Thus in this larger motorboat expected to be our home for another eight days we had a total of 12 people. Both the tour guide and cook also functioned as the boat hands while the mechanic also functioned as the boat captain. At about 4:00 pm that evening as the boat propelled itself on the river the tour guide and the cook started to put up the hammocks and the mosquito "nets" over the hammocks. The mosquito nets were fastened to the floor of the motorboat to prevent mosquitoes from our sleeping areas. Although we were on weekly anti-malaria medicine, no chances were taken deep in these jungles. As with Rio Juruena on this trip, there was nothing to see other than the thick-wall of tropical rainforest, the waters of the Rio Guapore and sky with blue or rain clouds overhead.

No germplasm collection was made on the first day as we traveled as late as the mechanic felt the boat could carry us up to the spot that had been identified by SUDHEVEA for seed and budwood collection. By 5:00 pm our cook had already gone into action to get dinner ready. Our boat anchored after almost twelve hours. There were no villages or huts in this area, so in pitch darkness using flashlights and lanterns our boat was secured with ropes and cords to one of the largest trees on the bank of the river. Then we settled to our meals as we listened to sounds of busy night creatures in this tropical rain forest. Knowing that we had a long working day ahead of us, we went to bed immediately after dinner. We all swung from our hammocks like beds lined up in a military campsite at night.

We were up at 5:00 am the next morning. Using the forest as bathroom that early in the morning brought memories of being across the Atlantic in Africa. I was fully aware of having escaped black mamba strikes while doing the same things in the Niger Delta. As scary as it may sound, we had no option. By 6:30 am we had taken our breakfast, picked our backpacks and started off on our trail. The cook and mechanic/captain of the boat were left on the boat. Thus, ten of us took off on the trail collecting seeds and budwood materials as could be found. After 15 kilometers we came to the compound of a seringuero and his family. We chatted with him for about ten minutes as our team leader, through an interpreter explained our mission including the use of hand signals to explain different points of our tasks. He assigned one of his sons (about ten years old) to work with us and show us wherever we wanted to go for rubber trees. I wondered how a ten year old would help us to accomplish our goals. But after one hour working with him my doubts evaporated. I should have known better. I lived with my grandparents until I was seven years old before joining my parents permanently. My grandfather was one of the biggest and most productive farmers in the village. During those seven years I was always on the farm with my grandfather and learned a lot about farming. Thus, as this young "man" led us to every rubber tree I could relate and felt more comfortable. And so deep into the forest we went.

Very soon we came to an area with a rubber tree standing in water that was knee deep. For another tree our young man started to swim. Our team leader called on him and asked if there wasn't another trail on an up-land dryer area given that we were in a flooded part of the forest with rubber trees. The young man simply smiled, nodded in the affirmative but kept swimming to the next rubber tree. Of course we had to go along with him to this tree. By now the water was approaching waist level. With all the snakes around, we called on the young "man" that this was going to be the last rubber tree which we were going to allow the tree climbers to climb. I made notes of the environmental and physical conditions of these entire collection areas. As we waded our way back to the dry trail, we felt very exhausted. But it was a wonderful experience, which prepared us for the rest of our Rio Guapore expedition. By the time we returned to the boat is was already 7:00 pm and we were only three quarters dried.

The next day we pulled out into the river and continued downstream to our next collection site. It was on this day that I found out why our team leader bought all those alcoholic beverages and cigarettes. Two hours after leaving our last stop, we came to a military camp on the Bolivian side of the Rio Guapore. As our boat pulled up into the landing area, it was obvious that we had been spotted and followed for more than thirty minutes. An individual who appeared to be the captain of the troops in this camp came and stood at the bank of the river. He ordered two of his soldiers to take the ropes and cords as they were thrown out by our tour guide and cook. After our boat had been anchored and secured, our team leader jumped out of the boat and shook hands with the captain. A few minutes later, he asked us to come out of the boat to meet with the captain. Within a short time, our team leader introduced all of us. Then as we went back into the boat not knowing what would happen next, our team leader and the captain went into the tent housing the officers. Our team leader emerged from the tent signaling to the cook to carry all the boxes of alcoholic beverages and cigarettes and take them to the boot. When our team leader returned to the boat, he looked at my Thai colleague and I, rolling his eyes as if to say, "Now you see why I bought all that liquor." Of course by now we fully understood especially after he told us the background of the soldiers, and the fact that our germplasm collection downstream over the next few days took place on the Bolivian side of the Rio Guapore. Furthermore, this was going to be our way back to Villa Bella. Thus, it was very important to establish cordial relationship with a military outpost and border patrol individuals.

After this stint with the military officer and soldiers, we continued our journey downstream. Over the next two days we collected seeds and budwoods under some of the heaviest rainfalls that we experienced during collection. At the end of the seventh day we were completely exhausted. Our team leader requested the mechanic/captain of our boat to check on supplies and the condition of the boat to see if we could continue for another two days. The report was not very good. We were low on food and most importantly we had enough motor oil and petrol to be on the river another three days. And since our return journey was going to be upstream we would need all the power the motorboat could provide. That evening our team leader called our regular nightly meeting and finally made the decision that we should return to Villa Bella. There was no need for an international team of scientists to be stranded in the middle of the Amazon Basin. Because of the strength of the currents, a proposed 12 hour run became 24 hours.

Then he cautioned us that because of the report he got from the military base camp on the Bolivian border, there were two locations upstream which had many seringueros and could be areas for fruitful collection of seeds and budwood materials. These two sites were called Nova York and Querbim, both of which were 24 and 14 hours respectively from Villa Bella.

The way he made the statement, it was obvious that we would stop at these two locations upstream for seed and budwood materials collection. So early the next morning, we went through our usual routine. The cook had breakfast ready, the mechanic checked the motor-boat engine and started it in order to be warm for the upstream journey. We used our bathrooms in the forest and shortly after an unusual early breakfast we pulled of the riverbank into the main body of water of Rio Guapore. Interestingly, we were in full anticipation at the upcoming stops. Seven hours into our upstream ride the boat was anchored on the eastern shores of Rio Guapore in Querbim on the Bolivian border. Putting on our collection gears we stepped out of the boat onto this very dark interior of the jungle. Just when the last tree climber was coming out of the boat, we heard the cook scream. We all wondered why. He pointed to a spot to our immediate right where this long python was gradually coiling in its usual hunting preparation. With no guns in the boat, using a machete to try to kill such a huge snake was considered very dangerous. So we all had to quickly slash our way to move further inland for the start of the rubber seed collection. Meanwhile, the mechanic and cook started moving the motor boat around to see if the propeller could be used to either kill or scare the python away from the area.

Using the map that our team leader obtained from the border post earlier, our team leader, Paulo, carefully directed us into an open space with about five huts. We heard voices as we approached the site. We had already covered about two kilometers to get to this location. As is always the case, we identified the hut that looked like the one where the chief or head of the village would be living in. It appeared the seringueiros were having a short break for breakfast after completing the first part of tree tapping for the morning as rubber tree tappers always do. We approached and greeted them in a very friendly manner. Their reaction to us was friendly as well. They were prepared to help us, but they had already completed the eastern estrada for the day but we have plenty of rubber trees on the western Estrada. So the seingueiros said "you can follow us for your seed and budwood collections." As soon as they finished eating they were ready to get into the estrada, and we followed. We saw a lot of rubber trees either individually or in clusters. It was really worth the stop. We started off from the small village on this Estrada at about 10:00 a.m. and did not get done until about 4:00 p.m. having already covered more the 10 kilometers combined. Recalling our experience earlier in the day, Paulo decided to halt the collection so that we could get into our boat before it got dark. When we returned to the small village with our newly found friends, seringueiros, we thanked them very much and offered them token gifts of extra hats and shirts that we had with us. In fact they heard our story from earlier in the day, two of them volunteered to accompany us to our boat. They told us that the shores of Querbim were recognized by the natives as an area filled with snakes and crocodiles. The area is rich in fresh water fish, birds, monkeys and other animals so that wildlife flourished in the area and hence an abundance of food source for this far reaches of the Amazon Basin. I thought to myself, the very thick vegetation may explain the richness in flora and fauna of Querbim. I almost asked one of them how they were able to survive in such an area. On second thought, I did not ask not knowing exactly how Paulo and our other Brazilian counterparts would feel considering I have been asking a lot of questions during this stretch of the expedition.

The mechanic and cook were smiling when we got back into the boat. We thanked the seringueiros who accompanied us to the boat and said goodbye to them. Both the mechanic told us that it would be better if we left that particular site and traveled for about an hour upstream before stopping for the night. Paulo said "it was already getting dark, how are you going to steer the boat safely." The mechanic replied with a satiric smile and said "Which do you prefer? To have a crocodile or python crawl into our boat for the night or be safe? Alternatively, do you want us to keep running the engine every one to two hours to scare off these animals and not have enough petrol to get back to Villa Bella? Remember we are already two days behind and still have about seventeen hours left." The cook himself quickly added "Also we are running very low on food and jokingly added that our next two lunches will have to come directly from the forest and river. I will be putting out some fishing hooks once we stop tonight." By now we were all gathered around the mechanic and cook for even as they spoke they kept looking over our shoulders to the river and the nearby jungle.

Without hesitation Paulo laughed very loud and said "So, what are we waiting for? It will be very difficult for any one to find us here in the middle of no-where. Everyone should get ready and lets go upstream for about two hours until we clear out of this thick jungle with its species richness. The kerosene lamps were lit and the hammocks came down and we all got ready to eat and sleep. We were so scared that nobody talked about showering in the river that night as we would have done after completing the day's assignment. The talk from the mechanic and cook also overshadowed the tremendous success we had in our seed collection for the day. After about two hours upstream and with flashlights the boat was anchored on the shores of what looked like a mini open space on the Brazilian side of Rio Guapore. I leave it to the reader to imagine how we felt for the rest of the night knowing fully well that we could fall prey to any one of the jungle dwellers especially when we were running low on every supply from petrol to food. So it was great joy when Paulo told us that we would start our journey the next day when it was clear daylight instead of the usual 5:00 a.m. wake up and bathroom use in the forest. With seventeen hours to Villa Bella and a stop in Nova York we were beginning to walk a tight rope. Yet the excitement of collecting germplasm in the Amazon Basin, kept us hopeful that we will complete our tasks and arrive back in Villa Bella safely.

Having used the forest as our bathroom we ate breakfast and pulled out into the river. Our boat was holding up pretty well. Our cook had already reeled some catfish and tilapia and a few other fish species. Thus we were almost assured of a full dinner. Hence lunch will definitely come from the forest. We steamed along and by 12:00 noon we were in Nova York, again, on the Bolivian side of Rio Guapore. Conscious of the species environment we were now dealing with, the mechanic steered the boat into an area he thought was safest for us to come down. We could see tree-jumping monkeys enjoying their meals high up on those trees and dropping the seeds as they did. Here was gene flow mediated by monkeys in real life. Paulo and the team had agreed that no matter how many trees we found in whatever number of estrada we will stop collection once again by 4:00 p.m. and cover at least two hours to bring us closer to Villa Bella. Fortunately for us, just after two kilometers, we came across a village that was much bigger than what we saw in Querbim. The seringueiros already completed tapping for the day. It was going to be difficult to find any of them to return to the estrada considering the tappers were always out as early as 5:30 a.m. Therefore it was a great relief when we were told by the chief seringueiro that they had collected seeds during seed harvesting season and will be glad to sell them to us. Paulo immediately had a quick meeting with the team to determine the feasibility of doing so and the scientific implications of doing so. Once we saw the seeds, those doubts and questions were quickly erased from our minds. They were some of the largest and most varied seeds we had collected since the start of the expedition. After some questions to the seringueiro as to how the seeds were collected, Paulo called the chief to the side and informed him that we would pay for as many seeds as they had. Personally, I felt that this was a blessing from God, as this was an unexpected gift. After purchasing the seeds, we were still interested in seeing some of the trees that produced these very nice looking seeds. We spent an hour on one of the Estrada and collected budwood materials. We were back in the boat as early as 3:00 p.m. While in Nova York we purchased some food items, so lunch and dinner were covered. We pulled out into the river at 4:00 p.m. with seeds and budwoods from Nova York. Again we spent the night in the shores of the unknown place in the forest. But we were now estimated to be only four to five hours away from Villa Bella. This meant we could take our time the next day to get all our items at least pre-prepared for arrival in Villa Bella. This is exactly what we did. Low in food, petrol and engine oil we made our final pullout from the Rio Guapore shores deep in the Amazon basin en route back to our secondary base camp.

We finally arrived in Villa Bella twelve days after our initial departure. To my amazement, there were lots of people, children and adults all along the bank of the river waving and smiling as we pulled into the port of Villa Bella. It was a very tiring twelve days on the waters of Rio Guapore. We had suddenly become the new great explorers and stars of this old and charming city. I observed very closely even with my tiredness, the reaction of the black folks in the crowd with the sense of seeing one of their own conducting this scientific expedition deep in the jungle of the Amazon basin. It will be an understatement to say that I was overwhelmed by this reaction. As soon as we checked into our hotels I went to bed immediately considering how tired we were.

The rays of sun coming through my window woke me up the following morning. I immediately got ready for breakfast and the regular morning briefings from our team leader. Because of the one-day delay we experienced on Rio Guapore, our flight from Villa Bella to Cuiaba had to be rescheduled. Therefore we had a whole day free to do whatever we wanted. I saw this as an opportunity to see more of Villa Bella. So, as soon as breakfast was over I called on my Thai teammate to accompany me to the city center, which was within walking distance from our hotel. As we moved along I heard very familiar music being played in stores and shopping areas. There were the Jackson's, ABBA, Boney M, and a lot of country music by Dolly Parton, Anne Murray, Don Williams and others. At some point I stopped and went into one of the stores from which I heard music playing. The storeowner was black and to my delight, he had a son who spoke English. So I asked him through his son, how come he is able to carry such an array of artists. He smiled and joked that as soon as the artist's release their albums in America or Europe we are able to get them here in Brazil either through Rio de Janeiro or through Sao Paulo.

Sometimes people make copies of the cassettes or eight tracks and send them to us. “Finally,” he said “We have no problems getting these tapes to sell especially if the music was good.” I rolled my eyes and smiled back at him and told him that it was very interesting to see how far music could travel. My teammate and I bought several cassettes and 8-tracks from him in appreciation of his patience. Then we returned to our hotel. I spent the rest of the day reading, sleeping and preparing for our return flight to Cuiaba the next day. At the airport, several groups of people were there to see us off. I had already made some friends at the local hotel. Some of the ypung black children came to see me in the hotel when they heard we had an extra day in Villa Bella. Already my hotel manager was very friendly with us and we had begun to speak in terms of coming back to Villa Bella at some point in the future. Officials in the town while curious about what we had come to do were very happy to see a team of international scientists in their city which seldom made the Brazilian news headline. Therefore it was not surprising to see all of these senior officials at the airport, waving goodbye to us. I felt very fulfilled not only from the scientific standpoint but also from the socio-cultural perspective. I had a better and deeper understanding of Brazilian political system, culture and of course a tremendous knowledge of another indirect tributary of the great Amazonas. It turned out that Villa Bella would be the last far reaches of the Amazon Basin for my team to collect germplasm. Our flight from Villa Bella unlike that from Conomali was uneventful. We arrived safely and checked into our usual hotel in the city cente

# CHAPTER EIGHT

## JOSE FARIAS, BARRA DOS BUGRES: COLLECTION FROM THE OUTSKIRTS OF CUIABA

Within 24 hours of our arriving in Cuiaba in preparation for a return to Manaus, Paulo, our team leader, called a very quick meeting that evening. We were told that there was an area known as Jose Farias (Barra dos Bugres), about 200 kilometers (120 miles) from Cuiaba. It was said that even though the area was made up mostly of plantations there were still some wild rubber trees close to the nearby mountain from which rubber seeds could be collected. Next day we had breakfast at 6:00 am. Before we finished breakfast the SUDHEVEA already sent a vehicle with a driver and one technician to assist us with the collection. They were waiting for us outside the hotel. Paulo had reassembled some of our field equipments. The driver and the technician quickly loaded these into the four-wheel-drive Nissan Pathfinder sports utility vehicle. By 7:00 a.m. we were on our way. As scientists who worked on rubber we were fully aware of the fact that rubber tapping started as early as 5:30 a.m. a little before full sunrise and is often completed by 10:00 a.m. So it was very important that we got to the plantations in the middle of the active tapping period for the day.

Both driver and technician greeted us warmly and opened the vehicle doors for us. Paulo asked them if they had had breakfast. Both of them nodded in the affirmative. As we drove through the city and then to its suburbs and on to the main highway, it occurred to me that we had spent nearly three months and almost every weekend of that time here in Cuiaba and had not taking the time to learn a little history, geography, culture and economy of the town that had served as our Primary Base Camp in Mato Grosso. I knew Paulo did not know much about the town having come from a different Brazilian State. I could easily have picked up one of the flyers at the hotel and read about Cuiaba. But I felt a native would have a better knowledge of a place where he or she grew up. Therefore, I asked the driver, Carlos, what he knew about Cuiaba. Without hesitation he pointed to Luis, the technician, saying that he was born and raised in Cuiaba having obtained all of his primary, secondary and vocational education here in Cuiaba. I was very excited as I turned to Luis to give me a first hand account of his native town. Luis looked at me, smiled and started "Cuiaba our capital city of Mato Grosso, is a name of South American Indian origin that means "arrow fishing" reflecting the tradition of fishing with arrows. Some say the name comes from an Indian tribe known as Ykuipa. There is also a legend which says that a Portuguese individual was taking a bath in the nearby river with a plate made from coconut, called cuia, and the river carried it away. This prompted him to scream "Cuia Ba" meaning the "cuia is gone." Historically, the town was found around 1727 by one Rodrigo Cesar de Menezes, who came from Sao Paulo in search of gold. From a small town of less than 50,000 in the 1950s it has grown to nearly 250,000 inhabitants in the last twenty years. Culturally, Cuiaba has a very local culture that is based on Indian, African and Portuguese influences. It is home to an interesting Indian influenced native dances, cuisine, craftwork and music. Many of the local dances are also connected to the worship of Catholic saints, such as St. Benedict who is said to be the patron saint of the city. As you know, Carnaval is also dutifully observed for four days in this town. It does have its off-season Carnival called "Carnaval a de epoca or Micarecula." He concluded and smiled and asked if we wanted to know more about Cuiaba.

My immediate response was "Yes, Luis. This is a very fascinating story of your town. We are going to a place not too far from the city to collect natural rubber germplasm. Please describe the geography, climate, vegetation and even the economy of the city and its neighboring towns. You don't have to give any details." I saw his eyes brighten up as if to say I am very proud to talk about my city to this young African Scientist. I also observed Paulo showing significant interest in the discussion. So, Luis resumed his very proud description of Cuiaba. "Cuiaba, is the geographic center of South America and famous for being one of the hottest cities in Brazil with temperatures often above 40 degree Celsius with an average of 23 degrees Celsius. It sits in the transitional zone of the Brazilian ecosystems of the Amazonia, Cerrado and Pantanal. It is also close to a Mountain range called the Chapada dos Guimaraes (very near where you will be making your collections this morning). The city is often referred to as the Southern Gate to the Amazon. The commercial and industrial sectors of the city are based on agribusiness especially food processing. Mato Grosso is one of the most agriculturally focused states in Brazil and this is reflected in the overall business activities in Cuaiaba, the state capital. As you can see SUDHEVEA has an active office here." With that he finished his description of Cuiaba. Just then the driver, Carlos, informed us that we were approaching the plantations at Jose Farias. I thanked Luis for the tremendous education regarding Cuiaba. He nodded in response with a sense of pride.

Our vehicle pulled into the gate of the plantation. The gate opened and we were led to a small field office of the company that owned the plantations in this area. Seated to the southern corner of the office was Perreira Claudio. I figured that this gentleman was proudly managing the plantation. He welcomed us profusely telling us how he effectively managed the rubber plantation. The plantation was regularly monitored for the South American Leaf Blight to ensure that none of their trees succumbed to the deadly disease. Within forty minutes our arrival, we were in side the plantations surveying and identifying trees from which we planned to collect seeds and budwood materials. We wanted to be sure that we were not collecting seeds of rubber clones that had already been planted in the IRRDB member countries. So we were very pleased to see that the numbering reflected primary selections made by the Brazilian natural rubber breeders at the CNPSD. The rubber clones in each block revealed a unique numbering which became very useful once we started collection. Because this was a very large plantation we had to establish estrada-like approach as though we were inside the jungle along the Rio Guapore or Rio Juruena. With no tree climbers to help us with budwoods high up on the younger branches, our technician from SUDHEVEA devised a means of climbing the trees with a matchette hanging down his waist for cutting the required branches. He had worked at a rubber plantation before, so he knew how to get to these tree branches.

Using our newly established "estrada", we divided this plantation that was nearly eighty acres in size into about three long estrada. Because these were relatively improved rubber clones from primary selection, seed size, color and patterns varied but not nearly the amount of variation observed say along Rio Guapore or Rio Juruena and even in Arcatuba/Itauba locations. Yet there was sufficient variation that would reflect the first generation of a primary selection program. Mr. Claudio accompanied us and watched with curiosity as to the work we were doing picking up seeds, examining them, taking notes and carefully bagging the seeds for shipping to Manaus. His greatest appreciation came when, after we had selected the trees from which to obtain budwood, he watched Luis climb to the very top of one of the trees to cut the branches. Earlier, we had spoken to a number of the seringueiros about the yields of the clones/trees they tapped on a regular basis. Mr. Claudio asked a number of questions as to the ultimate objective of bringing rubber scientists from all the IRRDB member countries to Brazil for such an expedition. Paulo, our team leader explained to him, Claudio, in detail our objectives and the potential implications for the natural rubber industry if this expedition was not done. I could see Claudio listening attentively and wide-eyed. Just before finishing his explanation, Paulo asked Claudio if he knew of any rubber trees growing in the nearby mountain so that we could make collections and see if there was any possibilities for cross pollination between the trees in the wild and the rubber plantation. Claudio, put his hand in his chin in a pensive mood and after looking up he said "I am not sure, but I have heard the seringueiros say that there were some wild rubber trees up on the mountain trail about 25 to 30 kilometers (15-20) miles from here. I have never gone up there. Considering the fact that you and your fellow scientists, have been known to go into the far reaches of the Amazon Basin, this should be relatively easy for your team to handle." Claudio sarcastically concluded his statement with a smile. Paulo thanked Claudio, turned around and looked at us as if to say "Be prepared, we are heading for the mountain." Over the past three months I have come to know this particular look from Paulo and no doubt it was definitely a signal for us to wrap up collection at the plantation and head for the mountain. He conferred with us and we all agreed and looked forward to the excitement of spending one or two hours going up the mountain. We thanked Claudio and all the seringueiros, who had by now completed their tasks for the day.

We wrapped up the collections (seeds and budwood materials) around the rubber plantation, and prepared for the mountain climb. It was already 11:00 a.m. when we left the plantation. The SUDHEVEA Nissan Pathfinder was a four-wheel drive so we did not have any problems arriving at the starting point just at the foot of the mountain nearly thirty kilometers from the plantation. Using a compass and a map of the area that Claudio had given to Paulo we mapped our estrada. We were fortunate to have Luis with us, for having grown up in Cuiaba, he recalled coming to the mountain as a young boy.

Thus as we put the estrada together, he worked very closely pointing out possible spots where we could find wild rubber trees. Once ready, we drove up about a 100 meters (300 feet) and saw the first rubber tree. The driver, Carlos put on the four-wheel function of the vehicle and turned slightly right to get us to the first rubber tree. We came down and walked up to the tree. As we circled around the tree we made notes and picked up seeds checking them for diseased and broken/cracked ones. Not very far away to our right from the tree was a cluster of three or more natural rubber trees. We collected more seeds and Luis our technician was able to get some budwoods. We remained in this area for another four hours going as high as 300 meters (600 feet) above sea level. We thoroughly enjoyed the exhilarating drive and walk through the forest on one of the few mountains we traversed during the Germplasm 81 expedition in Mato Grosso. The excitement level was so high that we basically did not stop for lunch. Therefore, as soon as we made our last collections, we drove to the restaurant about 100 kilometers towards Cuiaba. Because all the shipment supplies were already exhausted we had to stop at a supply store to buy Styrofoam containers, and polythene bags to convey the seeds and budwood to Manaus.

It was 6:30 p.m. by the time we got to our hotel. Luis and Carlos volunteered to help us package all the materials before they left thus ensuring that this last minute effort did not go to waste. We thanked them and invited them to a drink, which they cordially refused. Luis indicated that they were already doing overtime and were not sure of being paid. But they felt that this was a very important activity not only for IRRDB but also for Brazil. Carlos added that we had already fed them earlier as we returned from the mountain and Jose Farrias Plantation. This was the first and only collection combing a plantation and mountain rubber trees so even though the Jose Farias collection was unscheduled it turned out to be quite productive and a nice way to conclude the physical collection exercise of Germplasm 81.

Paulo and I went into our hotel room and for the first time both of us decided to say a brief prayer together thanking God for all His many blessings having just completed this major task on behalf of the world's natural rubber industry. We had our showers and went down to the hotel bar for snacks and some soft drinks. Once we got back to our room we immediately went to bed for the night.

# CHAPTER NINE

## FINAL DAYS IN CUIABA

Prior to leaving Mato Grosso we spent three days and two nights in Cuiaba to recap all our efforts and experiences in this Rain Forest-Savannah transition zone with its cerados, mountains and heavily thick rain forests in the far recesses of the Amazon Basin. Our team leader assembled us together on this bright Saturday morning in April, to take stock after a very nice breakfast. We were scheduled to leave for Manaus the following Monday afternoon. By now we had already bid farewell to our tree climbers and cooks. In fact, it was the scientific team *per se* which undertook the last collection at Jose Farrias. We had all our field notes and records reviewed to ensure that we had all documented the actual seed and budwood counts which we sent to Manaus. We suddenly realized that we had collected 8,908 seeds and 595 meters of rubber budwood materials. The results were far in excess of what we expected to accomplish. Most importantly, we recorded tremendous genetic variation among the germplasm we had collected.

I spent the second day visiting friends whom I had made in Cuiaba, among whom was Sophia at the photo store where I always processed the films of the pictures I took in the jungle. It was an emotional day when as we both reflected on that night when the rain thoroughly soaked us even as we prepared to share a wonderful night following a very delicious dinner. That night our team came into Cuiaba for our regular weekend rest and relaxation (R&R). Continuous visits to the photo store had brought Sophia and I close together and wondered when we could be alone together. Sophia was probably about 24 years old (I never did ask her age throughout all our meetings). I recall the manager, a young woman herself, probably in her mid-thirties, of the store once telling us in Portuguese to find the time during one of those weekend R&R to spend some quiet times alone. It was as though she could sense the feelings developing between Sophia and I. I never did understand given my poor knowledge of Portuguese. I believe Sophia was too shy to let me know what the manager meant or that she was very afraid of developing a relationship with a customer from such a foreign land especially an African. So here we were enjoying the ambience of the city. We were in one of the best restaurants in Cuiaba. It was well furnished with rounded-hut-like structures that made it very welcoming and romantic. We ate, smiled, joked about Mato Grosso, the jungle and of course Cuiaba. The music was very good and we danced every time the opportunity presented itself.

We talked about what she did at the photo store and what future she was carving for herself. Turned out Sophia was in her final year at the local university studying international relations and was very interested in interacting with foreign individuals in order to broaden her knowledge of the world. The store seemed to have been a great place for such practical education. But she had never expressed these emotional feelings with any one before, given her circumstances. She always maintained a strictly business relation with all her foreign customers. I supposed she figured at 29 years of age, I was not too old for her. Furthermore to see some one as young as I being a black person, and mingling with the top scientists and officials in Brazil, meant that I had to have something going for me. Whatever the case may be, we were getting emotionally involved. She worked at the store to earn additional money to compliment her scholarship and help her parents who had only a modest living. The store manager had been extremely nice to her by providing needed hours and the flexibility to work when she could. It had been a very wonderful experience for her. In about six months she will be graduating and looking forward to working in the Foreign Affairs Ministry. I suggested graduate school to her. I told her that I am starting my Ph.D. in Forest Genetics and Tree Breeding at Iowa State University, in the fall of 1981. And that perhaps with a M.S. degree she could earn much more than with a B.S. degree or rise faster in Foreign Service career, and even become an ambassador for Brazil in the future. She could even go on to earn a Ph.D. in International Relations and teach at a leading Brazilian University while serving as a consultant to the Brazilian government. She smiled and insisted that she had to support her family especially her junior brother who had just been admitted to a prestigious university in Brasilia.

We continued our discussion oblivious to all that was happening around us. When we returned from the dance floor the next time around, we arranged our chairs so that we could sit side by side rather than across from each other. She pulled out her wallet and showed me a picture of her family including her immediate junior brother whom she was very fond of and now very proud of because of his recent accomplishment. Then she looked straight into my eyes and asked a series of questions which I had anticipated given the fact that we were now pressing very hard on each other's body as we moved closer and closer to each other with every question and conversation.

"Do you have a family? Are your parents still alive? Are there any brothers and sisters? Do you have a girl friend or are you married?"

Continuing, Sophia softly said "I believe a young man like you with such great potential and having attained this level of international experience must have someone who is already dear to you." "Is that true?" she concluded and gave me this smile that almost melted me away. Deep down, I realized I was been swept off my feet and that if I did not do something quickly my emotions and human weakness would take over and the end result may not be very good from my well cultivated catholic and altar boy image in the circles of my family and friends. Yet, here I was with a very beautiful young Brazilian lady about five years my junior and from our conversation so far as ambitious and family oriented as I am. It is as though we shared similar things and thoughts at that moment. I have truly enjoyed her company over these past three months coming back from the jungle every weekend and always looking forward to going to the photo store that Saturday morning to drop-off my rolls of Kodak films in full anticipation of seeing and chatting with Sophia for a few minutes or as long as the store manager permitted her to do so. It was a cosmic relief from all the trekking in the jungle.

I was going through these thoughts and how best to answer Sophia's questions when the waitress at the restaurant came to our table and asked if we needed more drinks. I turned to Sophia and asked her if she needed any more drinks. Then I noticed that she had dropped her face and seemed to be in deep thoughts herself. I looked up at the waitress who smiled and said "Is there any problem with her or do you want me to come back?" Confused about what was suddenly happening around me I sheepishly replied "Yes, please come back later." Then I returned to Sophia, who by this time had not only removed her hands from mine but also gently pulled her chair slightly away from my chair. As I looked at her I observed that she watched the waitress walking back to the counter. It was not such a good look. Upon reflection, I asked myself if there was anything between these two young ladies that I did not know about. Of course I am a foreign scientist in Cuiaba and knew nothing about the local community. It is possible that both of these ladies knew each other very well, or Sophia may have been leading me on and I was going along, and here was a friend of hers who did not want her to get into any commitment which may not be good for her. It is possible that the waitress must have been watching Sophia and I engaged in conversations and sharing a body language which did not sit very well with her. And while one may not have liked her approach of coming directly to our table, staring at Sophia and asking me if we needed more drinks in a way that made Sophia feel uncomfortable, it woke me up and my conscience was thereby pricked. Sophia's reaction to the waitress's question and body language also gave me an immediate food for thought.

Thus, instead of answering her previous questions, I found myself asking her if she knew who the waitress was and why she suddenly went from what seemed like an emotional high to an emotional low. "Sophia," I softly called. "Do you know the waitress? I am surprised about your sudden withdrawal and emotional recoil as you reacted to her presence." She quietly replied "Daniela and I used to be very good friends having attended the same secondary school. Our families were close. But as we grew up the usual things started to happen. Daniela could not go to the university at the same time as I did. When I came home for holidays one year, I met her working in the same photo store where I am now working. Within a week the store manager fired her and hired me. None of us knew why she did it. In the meantime word had spread that her former boyfriend and I were dating. That was not true. After we all graduated from secondary school, her boyfriend, Barbosa, moved to the South to attend a university there. He did not bother to continue the relationship. Then lately, people from our community started to talk about you and me. It is said that Sophia who made the store manager fire Daniela is now befriending this young handsome scientist from Africa. Looks like Daniela, has been looking forward to seeing me and my new scientist boyfriend. So this was an opportunity for her to react to me, her one time friend who was now taking everything away from her."

Having now recovered from my emotional train, I sat up straight and gently patted Sophia on the back to reassure her that there was nothing to worry about. I reminded her that she had asked me some very interesting questions which I had to answer so that she could have a peace of mind and not be as disappointed as Daniela after my departure from Cuiaba. She apologized for her sudden reaction to Daniela indicating that she was still afraid of her not knowing how she has taken all that had happened to her through no fault of Sophia's. With a little smile coming out of her and the fact that both of us were out of our emotional trance, I felt the time was right to answer her initial questions. "Sophia," I called as I started to answer those questions. "Yes, I am married to a very beautiful woman, maybe more beautiful than you (she looked at me and smiled again). God has blessed us with two sons. We have been married for four years now. My mother is still alive and helping my wife to take care of our children. I have seven brothers and two sisters and I am the most senior among us. We lost our father to a ghastly car accident about four years ago, shortly after he retired from the Nigerian Federal Ministry of Transport, Inland Waterways Department. I had been working as Research Scientist in- training soon after my National Youth Service Corps when he retired. We had a lot of plans but the accident came suddenly." Sophia, noticed that I was getting quite emotional for a different reason and she quickly handed me a tissue paper that was lying on the table. Then it was her turn to pat me on the back and console me. Finally, we were both sober, and just then, Daniela returned and with a broad smile and once again asked if we were ready to order our drinks. She put up an air of accomplishment. It was as though she had been observing what transpired between Sophia and I after her initial interruption. Sensing this situation, my response was that we would prefer she asked someone else to serve us. As she turned to leave us, I noticed that Sophia and her looked at each other and shook their heads. Daniela could be seen moving away in walking manner of saying "I finally got her!!"

A new waitress was sent to us and we ordered our food and soft drinks along with water. The excitement of the evening had already left us but we managed to talk about what the future held for both of us moving forward. I reminded Sophia that she was a college student soon to graduate and join the rest of the working world. And as for me, I will be returning to Africa to prepare and travel to the United States to study for my Ph.D. in Forest Genetics. The look in her eyes indicated that Sophia had now fully revived from the emotional let-down as we talked about the future.

"Yes, Kome, I am very excited about graduating soon from the university. In fact meeting you is a sign that my degree in international relations will become very handy." Sophia said as she was finally able to look at me once again after several minutes of facing down on the table. Continuing, she asked "By the way, what is the time now? I have to be at the photo store very early on Monday morning. Meanwhile there are assignments which I have to finish after mass tomorrow." Then I told her that it was 9:45 p.m. and that we could finish our meal and then retire for the evening. "By the way, I am very pleased with your commitment to succeed. Your future looks very bright." I replied to Sophia. Suddenly, I realized that we were discussing things which both of us eldest children in our families liked to talk about. At this point I called on our new waitress to give us the bill for our food and drinks. I felt satisfied to make the payments and to see Sophia smiling again.

We stepped out of the restaurant. I felt a cold breeze and everywhere looked very dark even with Cuiaba street lights. I immediately knew what was about to happen as I held Sophia's hand to walk her home for the night. In the tropical rain forest, these cold breezes with the corresponding very low dark clouds were signals for an impending thunderstorm. Even before I could finish my thoughts, we saw the first lightning flash right before us and almost immediately a heavy down pour started. With no umbrellas and about thirty feet away from the restaurant, we had no choice but to return to the restaurant itself. We were drenched by the time we walked in. We sat at the very far end consoling ourselves. Personally, deep down I reflected on my tribal traditions and made a joke: "Sophia, my wife is reminding me through this rainfall and our being thoroughly soaked that I should not have taken a beautiful girl like you out even though I am six thousand miles away from her and our children." I made the statement to keep up our spirits. Sophia replied "That is very interesting. You mean, she sent a rain cloud after us. What if I said it was Daniela who was making her revenge on me, what would you say?" Sophia responded. We both laughed and waited patiently for the rain to stop.

Characteristically, the downpour stopped within fifteen minutes of starting. In our wet clothes I walked with Sophia to her home which was five blocks from the restaurant in the same direction as my hotel. As soon as her mother opened the door and let us in, I waved to and said good night in my newly acquired broken Portuguese. She waved back and her mother closed the door. I walked very briskly back to my hotel knowing fully well that tropical rains often come in bands in a sequential manner. I wanted to make sure I did not get fully drenched for a second time in one night. Several thoughts went through my mind and wondered what could have happened if Daniela had not shown up at the moment she did. "Was I losing my innocence and honest boy image in a foreign land or was I just trying to be nice to someone who had helped me with my photographs and who also demonstrated an ambitious intellectual future like myself." Then it dawned on me that the Lord Almighty had just rescued me from a very difficult situation. I made up my mind to attend mass the next day.

I opened my room door at the hotel. Seated on his bed and watching television, was my Brazilian team leader, Paulo. " Where have you been? How come you are so wet? You know as a member of this team you cannot be out there by yourself for long periods of time. You have got me worried." Paulo concluded. Still soaked and wondering how to respond to Paulo, I went straight to the bathroom to remove my clothes.

Once I felt comfortable in my new clothes, I apologized to Paulo for coming back to the hotel late. I told him that Sophia and I just wanted to have a quick dinner to give me a chance to thank her for all the help she had given to a foreign scientist like me. Paulo winked his eyes and said jokingly "You were trying to enjoy the fringe benefits of Germplasm 81" expedition. I can't blame you being so far away from home." He concluded with a smile. I quickly said "Paulo not like that. We just had dinner and stepped into the rain before we knew it, and then got drenched." We both laughed at the satire.

"You know we are scheduled to leave Cuiaba at 2:00 p.m. on Monday after meeting with SUDHEVEA officials here in Mato Grosso in the morning." Paulo said.

"Yes, I am aware of the proposed return to Manaus for a final meeting with all the expedition members." was my reply. Continuing, I said "Paulo, I'd like to attend mass in the morning. Will you be able to direct me or have someone take me to the church."

"No problem, I will." "Thank you" I said.

I had already started to feel sleepy. So I wished Paulo good night and went to bed. The next day, Sunday morning, I woke early. Paulo had already spoken with the hotel front desk and an attendant was asked to take me to this very beautiful cathedral in the center of Cuiaba. And because our hotel was in the city center, almost everything we needed was within walking distance. This was also true for the church. The mass went very well. However because I had not been to confession, I could not take communion on that Sunday. But I did feel a sense of spiritual relief after what happened the previous night from the perspective of a former altar boy. I returned to my hotel and went to sleep after a very delicious lunch.

It is Monday morning of April 13, 1981. We have packed up all our expedition gears and belongings ready for departure. Paulo and our Portuguese translator were already at the hotel counter checking us out. Breakfast was very delicious. Our meeting with the SUDHEVEA officials would take place at 11:00 a.m. "There were still 90 minutes before they come to take us to their offices." I thought to myself. "The photo store is only 10 minutes away and I think I can get there, see Sophia and Julia, the store manager one more time before leaving Cuiaba." I concluded. So, I asked Paulo for a thirty minutes permission to dash to the photo store. He looked at me with a wink in his eyes and said "Kome, I know-----. Okay, go and please return on time for us to meet our SUDHEVEA officials." "Thank you, thank you!!!" I said to him as I turned around and walked very briskly to the photo store.

Sophia was in her usual corner attending to a customer. As soon as I walked in she looked up and our eyes met. I could see her quickly completing her task with the customer and motioning for me to wait. I understood and stepped to the side. Just then, the store manager, Julia came out of her office. "Kome, good morning. How are you doing? Sophia told me all that happened last night. She also said that you are leaving Cuiaba for Manaus today." She said smiling as she walked towards me. I thanked her immensely for giving Sophia some free moments to chat with me whenever I came to the store. Then she looked around and saw Sophia busy with a customer. "Let me go get her for you. I will finish what is she doing. Well, it has been nice knowing you. Come again to Cuiaba whenever you make it to Brazil." Shortly thereafter, Sophia came to where I was standing. We hugged each other and then exchanged looks as if it to say, "It could have been." We stepped out of the store while waving goodbye to Julia. Then Sophia and I spent the next five minutes chatting and laughing and recalling the events of the previous night. She said her mother had become worried about her that evening. But in the end everything was all right. After much reminiscing we hugged each other again and held our hands for about a minute refusing to let go. "Sophia," I said "My colleagues are waiting for me and as I told you last night we are going back to Manaus today. It has been very nice knowing you and who knows our paths may cross again. This is a very small world. You are a very young lady with a clear determination to succeed. Continue to work hard and stay focused. I hope you will go to graduate school after obtaining your first degree. Greet your family and God Bless." I concluded my statement. I could see that her eyes were beginning water and she turned away briefly from me to wipe off the tears. Then she said softly "Kome, you have been a great inspiration for me. I have never enjoyed my work in the photo store this much. Through your pictures I have seen other parts of Mato Grosso. Most importantly you have opened my eyes to see that a young black man can be an international scientist. We always had this perception that it will be difficult for young Africans like you to attain such levels. The stereotype here is that young black boys and men are only good at playing soccer. You have changed all that. You know Julia likes you very much. That is the only reason she let me talk with you in the first place. I have heard all that you said and I will do my best to succeed and help my family by the grace of God. Your wife must be a very lucky woman to have a husband like you. Please do not forget me when you return to Africa, I mean Nigeria. Have a safe journey back to Manaus and to Canada and all the way back to Benin City." As she concluded her comments we hugged one last time and waved goodbye to each other. I turned and walked back slowly to the hotel reflecting upon the fact that I had made such wonderful friends in Mato Grosso in such a short time similar to the friendships I had cultivated during my National Youth Service Corps in Yandev/Gboko, Benue State of Nigeria several years earlier. These are the human aspects of scientific expeditions which are often not reported in scientific journals.

The vehicles from the offices of SUDHEVEA were already packed at the hotel when I returned. Drivers had been sent to bring us to the offices of the officials. Paulo took a final count of what was left of our team and we thanked all the hotel staff for their hospitality over the three months. The officials of SUDHEVEA and the Ministry of Agriculture and Natural Resources warmly welcomed us. Paulo gave a brief account of all the activities we undertook in the various jungle including unique experiences in Conomali and Villa Bella. He specifically thanked SUDHEVEA for the tremendous logistical support they provided to our secondary base camps. In reply, the officials indicated that they were very pleased with what they had heard about the efforts of our expedition team in Mato Grosso. It was concluded that there was great anticipation for expanding the genetic base of rubber clones in plantations for many years to come. After the exchange of pleasantries between the officials and our IRRDB expedition teams, the meeting ended at about 12:30 p.m. on that day.

We boarded an 18 passenger Brazilian made aircraft for Manaus. Having ridden in the small twin-engine plane continuously for nearly three months this plane appeared larger than it would be ordinarily. Surprisingly, all the officials of the SUDHEVEA organization with whom we had the meeting accompanied us to the Cuiaba Regional Airport. I felt very honored by this kind action on their part. We waved to everyone as we walked on the tarmac and climbed onto the plane. Keeping to schedule, the plane started its take-off on the runway at exactly 2:00 p.m. local time. I was fortunate to have a window seat which enabled me to view the city of Cuiaba during take-off and the rain forest during the three hours flight to Manaus. Flying over the carpet-like over-story of the tropical rainforest after spending nearly three months inside the jungle *per se* looked very overwhelming. It was a commercial flight and for some reason we as members of the germplasm collection team sat on seats far apart and did not give us any opportunities to reflect on what we had just accomplished. Be that as it may I sat and read a copy of the Newsweek magazine I bought at the Cuiaba airport. There was no major incident on this flight arriving safely in Manaus as the sun was setting over the River Amazon in the State of Amazonas.

# CHAPTER TEN

## REUNION IN MANAUS

Beginning with the Rhondonia team, the Germplasm 81 group started to reassemble in Manaus at CNPD as we had all been instructed to do. Acre was the second team to arrive in Manaus, while my Mato Groso team was the last to arrive. To my amazement, a majority of the senior officials of EMBRAPA/CENARGEN who welcomed us in Brasilia had traveled to Manaus to receive us at the end of the expedition. Among these officials was a leading female scientist, Dr. Nathalia Gasol, a plant pathologist. The plan was for us to have an all day meeting in which each team would report on the efforts undertaken in their various states. So at 8:30 a.m. we assembled at the conference hall of the Oil Palm and Natural Rubber Research Institute (CNPD) for the final presentation. The Mato Grosso team had a short meeting the previous night to review the report we had drafted in Cuiaba. Paiva our team leader was the last to present. Just before he took his seat he made a one-liner joke. "Our African Scientist also collected or seem to have collected a much larger Brazilian germplasm in Cuiaba." There was a pleasant laughter from everyone. I quickly replied "Yes, but the germplasm was so large I could not move it from Cuiaba. Furthermore, my indigenous germplasm in Nigeria will not allow me to establish an exotic germplasm." The laughter grew louder until Dr. Nathalia Gasol, who was chairing and moderating the meeting waved her hands to restore quiet. Even then she added a line of hers "It is good to know that Mr. Onokpise is not taking a large Brazilian germplasm to West Africa to ensure that the gene pool stays here otherwise it would have been nice to move that germplasm far and wide. We may have to find new pathological approaches to resolve any hybrid blight problems." Once again, everyone laughed and she had to raise her hands for quiet. Then Dr. Gasol spoke as a representative of EMBRAPA and the Brazilian government. "On the basis of the reports we have received in Brasilia and listening to your presentations today, there is no doubt that the expedition has been very successful for both the Brazilian Government and IRRDB. It is my understanding that three to five acres have been set aside for propagating the seeds, budwoods and seedlings collected. This area is now called the Brazilian-IRRDB Natural Rubber Germplasm Nursery/Center. In keeping with international agreements on germplasm collection and the movement of plant materials, fifty percent of all materials (seeds, budwood and seedlings) collected will be retained here in Manaus while the other fifty percent will go to IRRDB institutes. Today we celebrate the hard work of everyone, from the research scientists at EMBRAPA and CERNAGEN to the officials in Sud-Hevea as well as the tree climbers and the many individuals you met in the remotest parts of our country. This has been a unique working relationship and partnership between one country and a very important international organization. Finally, I like to emphasize that with the establishment of the Brazilian Germplasm Collection Center here in Manaus we are confident that natural rubber plant materials will continue to be made available for the foreseeable future to enable our scientists to conduct research that will enhance and promote a sustainable natural rubber industry." She gave such a nice presentation that she was given an applause for nearly three minutes. It was very rewarding to hear a female scientist not only well accomplished her scientific profession but also an excellent representative of the Brazilian government.

Dr. Subramaniam, our overall IRRDB team leader then spoke on behalf of the IRRDB scientists and institutes. "In 1976 when the idea of broadening the genetic base of existing rubber clones was first conceived and brought to the attention of the chief executives of the IRRDB at their annual meeting in Indonesia, and the requirements spelt out, very little was known as to the implications of such an effort. That year marked the 100 years anniversary of Sir Henry Wickham 's collections. And in 1977 the IRRDB Workshop on International Collaboration in Hevea Breeding and the Collection and Establishment of Materials from the Neo-Tropics detailed all the approaches to be undertaken towards what you all have accomplished over these three months. Although, there was a preliminary mission to South America in 1978 as part of the recommendations of the 1977 workshop, the actual on the ground activities of Germplasm 81 will remain a great historical accomplishment for the IRRDB and its member institutes. You have not only represented the IRRDB well, you have been great ambassadors for your institutes and your countries. I like to take this opportunity to thank the Brazilian government for being such a generous host through its many agencies already mentioned by Dr. Nathalia Gasol. Attention to the minutest details made for smooth travels even when conditions seem extremely difficult. We hope that moving forward, Germplasm 81 will further strengthen the partnership as materials are now exchanged in a scientific manner without the old stereotypes of the development of a natural rubber industry in one region at the expense of another region. Simply put "the Brazilians will no longer accuse the Malaysians of stealing their rubber seeds." (There was laughter on this statement from everyone). I want to specially thank our Brazilian scientists and translators who served as team leaders for each of the three teams that went to Acre, Mato Grosso and Rhondonia. It was your individual commitment to working with our IRRDB scientists that enabled them to be effective in this once in a life-time experience. Let me also add that the fifty percent of materials going to IRRDB member institutes will first be handled as follows. The Commonwealth Mycological Institute (CMI) in the United Kingdom UK) has agreed to provide inspections for the collected seeds and that these seeds will be kept at the Tun Abdul Razak Laboratory in UK, with supervision provided by phytopathologists from the CMI. Meanwhile the Tropical Fruit Research Station in Guadeloupe agreed to quarantine the budwood materials. The main IRRDB germplasm and multiplication center was going to be located at the Rubber Research Institute in Malaysia (RRIM). This center will receive seventy five percent of the fifty percent IRRDB share, while the second center to be located in Abidjan, Cote d'Ivoire, West Africa will receive the balance twenty five percent of the collections. It is expected that scientific information from these centers in Brazil, Malaysia and Cote d'Ivoire would become readily available to all IRRDB member institutes." Dr. Subramaniam completed his statement and asked all of us to stand up and give a big round of applause to all our Brazilian counter-parts. Then he told us to reassemble that evening for further instructions regarding the anticipated week of quarantine in a temperate country.

Meanwhile, the Director of CNPD, Dr. Vitor Ferrera who was hosting this meeting also spoke for about ten minutes and concluded by saying "my staff and I have worked to receive the collections as you have sent them to us from the base camps in the three states where you conducted the expedition. As Dr. Nathalia Gasol mentioned, we have established a nursery for all the accessions. Our Research Farm Associate Director, Dr. John Bruno will be taking us through the nursery tomorrow morning starting at 8:00 a.m.. I hope you will enjoy that visit. Also, tomorrow afternoon following a tour of the nursery and lunch, the laboratories here at CNPD will be opened for all those who want to tour them especially for the ones who did not get a chance to do so three months ago. This should be a nice way to conclude your scientific experience here in the Amazonas." He concluded his statement with a smile. We all smiled and welcomed the opportunity to see what had happened to all those efforts in the jungles. Considering the genetic variation, from seed color and size our Mato Grosso team observed, I looked forward with great anticipation to seeing how the nursery would look like with the thousands of plants growing from seeds and budwood material. It was 4:00 p.m. when the meeting and presentations ended. Dr. Nathalia Gasol greeted each one of us with a hand-shake as she wished us the very best in our professional careers. Then along with the other officials from EMBRABA, Dr. Vitor Ferrera saw them off to the parking lot where they took the CNPD official vehicle to the Manaus airport for a flight to Brasilia. We retired to our hotel rooms.

Dr. Subramaniam had asked us to assemble for dinner in the hotel restaurant that evening so that we could discuss plans of the one week quarantine and other items of interest to our member institutes. Thus, at about 6:30 p.m. we all got together for dinner. It was a nice Brazilian meal with fresh fish and shrimps. Included with the meal was a plate of ground and fried tapioca (*Manihot esculenta*). What in Nigeria would be called gari. So I asked my room mate from Thailand who spent his three months in state of Acre if he had observed that plate of manihot served with meals in Acre. He indicated that, that was the case. I came to the conclusion that it must have been a cultural thing. Like natural rubber, *Manihot* is also native to Brazil, and it would appear its use for food items must be widely acknowledged even in big star hotels. Food was served and we all ate to our fill. At 8:30 p.m., Dr. Subramaniam stood up for us to finish with our meals so we could listen attentively to him. Then he congratulated us again for a job well done. He thanked us once more for being good ambassadors not only of our individual member institutes and country, but also excellent IRRDB scientists and representatives. Now, our job is done. Given the devastating effect of the South American Leaf Blight, the IRRDB executives agreed during the planning stages that none of us should travel directly to our countries all of which are in the tropics except for South China immediately after completing the expedition. We have to spend a week in a temperate country to ensure that we do not spread this deadly South American Leaf Blight disease that has basically prevented Brazil from becoming a major world producer of natural rubber. Therefore, we shall all be spending a week in London, England. Arrangements have been made for your tickets to reflect the new travel plans." We all felt very happy as it offered us an opportunity to get some rest while at the same time satisfying our responsibilities to the world's natural rubber industry.

Just as we were leaving the dining area, it occurred to me that I could travel to Canada instead of Great Britain. I said to myself. "After all Canada was also a temperate country. I just left there only ten months ago. I have friends there and I will be better served meeting our family friends and taking care of a number of things while fulfilling my quarantine requirements. Furthermore, I did not know anyone in Britain and I could suddenly become bored instead of enjoying the downtime provided. Of course there could be things to see and reminice about. Yet the thought of returning to the University Guelph campus was too rewarding to give up the opportunity" So, I immediately ran back to Dr. Subramaniam to request a change of location from London, England to Guelph, Ontario, Canada. I joked with him that Canada was in fact not only colder but will have more snowfall than Britain at that time of the year. This will ensure that all spores that may be stuck to me and my clothing and other items will be quickly destroyed. Dr. Subramainiam asked me to give him time to sleep over my request and he would have an answer for me the next day.

We were on our way to the Germplasm Nursery when I walked up to Dr. Subramaniam and greeted him with all the respect in the world. Before I could ask him of the decision on my request, he smiled and said. "Mr. Onokpise, your request has been approved. I sent a telegraphic message to London last night immediately after you made your request. The response came back this morning and it was positive. The conditions are that your air ticket should not be too expensive and that you will receive the same amount of per diem for the one week like your counter parts going to London." "Thank you very much, Dr. Subramaniam for standing up for me. I am indeed very grateful. I will be able to manage myself with the equivalent amount of per diem while I am in Canada," I replied. With this issue settled I joined my colleagues at the Nursery. Dr. Vitor Ferrera was already waiting for us. He introduced us to Dr. Jon Bruno who then introduced us to the nursery and field staff. Starting with the Acre collections, Dr Josef Mundo walked us through rows and rows of young rubber seedling plants. There was tremendous variation in the growth of these seedlings especially when we came to the Mato Grosso collections. At that point Dr. Paiva was particularly interested in those accessions from Rio Juruena and Rio Guapore. It was obvious that Dr. Paiva wanted to see how the seedlings from the very remote and distant parts of Mato Grosso were performing in the nursery, considering how much difficulty we had in reaching some of the areas for collection. I was very pleased to see this tremendous phenotypic variation at these very early stages of growth. We all wondered how they would perform later as they mature and are transferred to the field. One of the striking observations was the relative high germination rates for the seeds, knowing fully well that Hevea seeds were a recalcitrant group of seeds. Dr. Bruno told us the budwood materials were still in the greenhouse and it would be a few months before they were transplanted to the Nursery. Upon the conclusion of the nursery visit, Dr. Subramaniam thanked both Dr. Josef and Dr. Bruno for such an educational tour. We were all reassured that our efforts will be fully appreciated by the natural rubber industry for a very, very long time. After the Brazil-IRRDB Germplasm Nursery/Center tour, Dr. Bruno gave us an overall view of the rest of the research plots at CNPD. It was lunch period by the time we finished the visit to the research fields.

During lunch, Dr. Subramaniam told us again that Dr. Vitor would be making the laboratories available for us to visit.

All the scientists and lab technicians will be present to welcome us. Then he said that the evening and all of the next day will be free for us to prepare ourselves for departure from Manaus to the different temperate countries we had chosen to spend our quarantine time. Therefore after lunch and a short nap I joined the group who had already started taking the tour of the laboratories. The outcome of the visit was very interesting in the sense that despite the South American Leaf Blight problem, the Brazilian scientists still worked very hard to ensure that Brazil remains a natural rubber producing country even at a minor level until the time when the disease problem is resolved. This was commendable on the part of EMBRAPA. At the end of the laboratory tour, we were given some light snacks and soft drinks.

We had a free evening to get ourselves ready for departure from Manaus. I sought out Mr. Ong Seng Huart, one of the Malaysian Rubber Tree Breeders. I had read some of his papers in the early days of my employment as we were developing the Plant Breeding Division at the Rubber Research Institute of Nigeria (RRIN). In 1977 I represented RRIN at the Workshop on Hevea Breeding and the Collection and Establishment of Materials from the Neo-Tropics. Following this workshop I undertook a one week tour of the Rubber Research Institute of Malaysia (RRIM). It was Mr. Huart who hosted me. He exposed me to all the activities at RRIM from the plant breeding to crop protection research programs at the main headquarters in Kuala Lumpur to the Experiment Station in Sungei Buloh where all the field plots were located. The research factory, technology center and the training school were located in Sungei Buloh. He also took me to the small-holder operations when he further explained to me how the natural rubber industry operated in Malaysia and how this made it possible for effective funding of RRIM. My tour that week enabled me to learn a lot from Mr. Huart. Both of us had not been able to interact as much as I had wanted during our stay in Brazil simply because he was on the team that went to the State of Acre while I went to Mato Grosso with my team. Therefore, this one evening was the only time we had to see each other. Once I found him during the dinnertime, I requested to have a few minutes with him.

Before I could say a word, Ong said "Kome, how have you enjoyed the germplasm expedition? I am sorry we have not had time to chat. I recall we were so busy getting informed in Brasilia coupled with the fact that we were all just arriving for the first time. Even here in Manaus, we had very little time to really speak to each other. I remember you mentioning that you will be going for postgraduate work when you visited us in Malaysia about four years ago. How is that coming?" I responded with the same theme. "Ong, you are quite correct we have not had time to really chat. First, let me thank you for all the support you gave me during my visit to your institute in 1977. As for the germplasm collection, this has been an unbelievable lifetime experience. I am very thankful to God for this opportunity. With regards to my postgraduate studies, I just obtained my M.Sc. degree last year from the University of Guelph in Canada. Meanwhile, I have several scholarship awards from my home state and our federal government to start my Ph.D. in Tree Breeding and Forest Genetics at Iowa State University in the USA. However, I also received a Commonwealth Scholarship award to study in Malaysia. As you know, the commonwealth scholarship is more comprehensive especially as it relates to someone with a family like me. I thought about studying at a university in Malaysia, I will conduct my research at RRIM and do a dissertation on a topic dealing with developing new natural rubber clones. What do you think about my doing my Ph.D. in Kuala Lumpur with all of you."

Ong took a very deep breath and looked away for about fifteen seconds before responding. "Kome, first and foremost, Congratulations for your scholarship awards especially the

Commonwealth Scholarship. In Malaysia we all strive to get that prestigious award. In fact, two of my colleagues and I will be leaving for Australia in two years on one of those Commonwealth Scholarships. I have to be frank and truthful with you. We are sending our young scientists to Australia, Britain and the United States for their advanced degrees, why would you want to come to Malaysia for your Ph.D. when you already have other scholarships to go to the USA? Although, prestigious I am quite sure you will be in a much stronger position at this time, if you earned your degree from a university in the West. No doubt, I will very much like for you to come to Kuala Lumpur, but I have just given you my personal opinion." I looked at him with some kind of surprise. I was leaning towards going to the United States, but the prestige of a Commonwealth Scholarship and its potential support for my family while doing research on natural rubber in Malaysia was quite tempting. Thus as I listened to Ong and his honesty, I immediately concluded that he was right. If the country that had the world's leading Rubber Research Institute was in fact sending its young scientists overseas, then what guarantee is there that I would receive the best education possible especially for my terminal degree. Turning to Ong as we both sipped our guarana, the Brazilian soft drink we were trying out for the first time, I said. "Ong, I like to thank you very much for painting the true picture for me. I agree with you that at this time it may just be best for me to reject the Commonwealth Scholarship and take advantage of my other scholarships. I am very grateful for your thoughtful approach to what is important to our future. Thanks again. Thereafter, we both reflected on the activities of the past three months. He talked about his experiences in Acre including the decision to stop for the Brazilian Carnival. I talked to him about my experience in Mato Grosso and the Carnival as well. We exchanged addresses and wished each other the best of luck in our professional careers.

Over the next 36 hours, we all prepared frantically for departure from Manaus. Our air tickets were all delivered to us as scheduled. Issues of entry visas were raised but nearly all the countries involved in the expedition except for Cote d'Ivoire and China were members of the Commonwealth and at the time commonwealth members did not require entry visas into member countries. Dr. Compangnon was going to France anyway to spend sometime while our Chinese colleague who was with the Mato Grosso team would be getting a visa arranged by the IRRDB to enter Britain. As for me, I was going to Canada, a commonwealth country. In addition my multiple entry visa into Canada was still valid. So I was prepared. On a very bright morning, Dr. Vitor Ferrera informed us that the CNPD van was ready to convey us to the Manaus International Airport. Many of the CNPD research scientists came to our hotel to wish us farewell. We all greeted one another entered the van and as the driver pulled away we waved and kept waving until we could not see the CNPD folks. Inside the van for the next 20 minutes as we drove from hotel to the airport one could see the exchange of addresses to the nice treatment we had received from our Brazilian hosts. The hospitality shown at all levels and the efficiency demonstrated were to be highly commended. Many of us felt that we would like to return to Brazil as soon as the opportunity presented itself. At the airport we tipped the driver, checked in our baggage. I boarded a Boeing 747 owned by the Brazilian airline, VARIG. My flight was scheduled to go to Miami in the U.S. where I would then take a connecting flight to Toronto, Canada. I ran quickly to say a final good bye to Mr. Ong Seng Huart and my Thai friend on the Mato Grosso team. They took the British Airways flying directly to Heathrow Airport in London. Once I took my seat on the plane, I started to reflect on the three months that had just past so fast with a lot of excitement. I thought to myself if I would ever have such an opportunity again and be exposed to so many nice people in the farthest reaches of the world. Our VARIG pilots and hostesses prepared us for a very smooth takeoff and no sooner were we in the air did I fall asleep.

# CHAPTER ELEVEN

# QUARANTINE IN CANADA

The VARIG airline touched down so smoothly that little did we feel the impact of a Boeing 747 landing after nearly five hours of flying. This was my very first time at the Miami International Airport. It was very big and looked very well maintained. There were very many people at the airport especially those of Hispanic decent. As I walked through the concourses and observed everything in sight, I kept asking why there were so many individuals of Hispanic decent in this one place outside of Spain. I recalled that there were not a lot of Spanish speaking folks in Brazil. It took just 20 minutes into my two-hours stop over in Miami to observe that this was a Spanish-speaking city. When I asked why there were so many people of Hispanic decent in this city, I was reminded that all of Florida used to be a colony of Spain and that today Miami is considered the commercial capital of Latin America where people from Central and South America come to do business, sometimes on a daily basis. This would explain why in some stores there were vendors who could not speak English fluently. This notwithstanding, it was a booming city which was fully reflected in the traffic flow at the airport.

I finally located the gate for the flight to Toronto, Canada. "It is Air Canada!!!" I said to myself. Two years of studying in Canada had made all things Canadian interesting and valuable to me for some reason at that point in my life. The boarding announcement was made and we checked into the plane and sat down to listen to the usual preflight instructions. The Flight Attendant spoke in English and then repeated the same instructions in French. I was beginning to wonder why she would repeat the instructions in French when it dawned on me that Canada was a bilingual country. As would be expected from a Canadian operated airline, everything was done to the book and the plane took off right on schedule. After three hours of flight in this Boeing 737 jet, we landed at the Toronto International Airport. I cleared immigration without any problems. I had my University of Guelph hat on and it was like a home coming for me. As I came out to the taxicab area with my luggage, I felt the cold breeze of April in Canada rush through my face and body. "What a change? I just left the humid and hot Amazon Basin, and here I am in the cold atmosphere of Canada. I am truly ready for QUARANTINE in a temperate country," I subconsciously told myself. After negotiating a reasonable price with one of the drivers I got into the taxi and headed for Guelph.

Once I saw the sign reading "Guelph, 20 kilometers" I started thinking about where I would stay. "Should I be close to campus or return to the area where my family and I lived in an apartment about a year ago?" Just when the cab driver turned into the Guelph exit, I concluded that I would stay in a hotel close to the Westwood apartments. As I recalled, the apartment was very close to a major mall, it had easy access to many roads and there were at least two bus stops within walking distance. So even before he asked, I told the cab driver to take me to the Holiday Inn near Westwood Apartments. Because, I could not remember the exact street name I directed him to the area through instincts and memory of some familiar buildings and other features. It was about 7:45 p.m. just as the sun was setting when we arrived at my hotel. I paid the driver and thanked him immensely for being patient with me. The lady at the hotel counter welcomed me with broad smiles and as we spoke she treated me with such warmth as if to say welcome back to Guelph and Canada.

I opened the door of my hotel room on the second and final floor of the hotel, threw my jacket and other clothing on the bed and simply laid flat on my back for the next fifteen minutes ruminating on all that had transpired in the past three months and here I was back in Guelph, ten months after my family and I skipped my graduation ceremonies to return to Nigeria in order to fulfill the conditions of the scholarship award from my employers, Rubber Research Institute of Nigeria. I was beginning to feel sleepy when my stomach reminded me that the hunger signs were pulling on it. Then I jumped up, used the bathroom and came out of my room to survey what was nearby for food. To my immediate left was the mall with all the possible restaurants, to my right were a string of fast food places, McDonald, Burger King, Wendy's and the others. Straight ahead about three-quarters of a mile away were the Westwood apartments, and just behind me was the road leading to Expressway 401 which runs north-south from Toronto to Windsor on the southern border of Ontario Province. "Yes!!! This is why I wanted to stay on this side of town for quarantine" was my thought process. "I have not had a burger since we returned to Nigeria. Here is my opportunity to do so." I thought. Without further contemplation, I went straight to McDonald. I ordered a complete meal and ate to my satisfaction. The French fries were particularly tasteful. As soon as I got back to my hotel, I went straight to bed since I was very exhausted and felt sleepy. I knew I had a full plate of activities for while here in Guelph.

My first port of call at about 9:00 a.m. the next day was the Imperial Bank of Commerce located on the University of Guelph campus. I had been carrying travelers' checks throughout the germplasm exercise in Brazil and could not change them while I was still in Brazil. As part of a per diem for the Brazilian travel, my employers, RRIN, had given me about six thousand five hundred dollars ($6,500.00). I converted $6,000.00 to American Express travelers' checks and kept $500.00 as cash before leaving Benin City. Upon arrival in Brasilia, I thought about changing the entire per diem into Brazilian cruzeros but did not do so for two reasons. First, following our initial briefing in which we were told that IRRDB was going to be fully responsible for all expenses (accommodation, meals and incidentals) throughout the period of the germplasm collection, which meant that we were not able to spend much of the personal funds we may have brought with us. Secondly, during one of our breaks, I had gone to one of the banks in Brasilia where through the aid of the hotel receptionist I wanted to obtain some local currency, cruzeros. The bank manager made me to understand that under the Brazilian regulations at the time for tourists, all monies once converted to cruzeros had to be fully spent in Brazil. Therefore, I changed only three hundred dollars into cruzeros. These were sufficient for all my personal purchases in Brazil.

Now in Guelph, I could fulfill all the commitments to my friends and get something for my family. My account at the bank was active and a number of the bank tellers still remembered me. I requested to see the manager who had been very nice to my wife and I in May of 1980 just before we returned to Nigeria after defending my M.S. thesis. He had granted us a small loan, one of our family friends, Brian and Kathy McBlain cosigned for us. We had arranged to remit the payment from Nigeria on a monthly basis. It was a great leap of faith and an unqualified trust on the part of the bank manager and our family friends. We kept our promise once I settled down on my job at RRIN. But it was going to take nearly three years to completely pay-off the loan at that rate. So the Amazonian expedition was a tremendous blessing especially that I was able to travel back to Canada and visit Guelph to resolve an issue like this and maintain the credibility which in the first place enabled the manager and my friend to do what they had done.

As soon as the bank manager saw me, he rose from his chair and motioned me to come in and have a seat and with a firm handshake along with a smile, he said. "Good morning, Mr. Onokpise. What brings you to Guelph? Brian told me that you and your family had already returned to Nigeria. Anyway, welcome back to Guelph. What can I do for you?" This was a very heart-warming reception for me from someone who took a leap of faith on me. I leaned back on my seat and said, "Good morning sir. Yes, my family and I actually returned to Nigeria last May without even participating in the graduation ceremonies because my employers wanted me back once I had completed my program. With air tickets and our luggage already sent, we had no choice but to return. It turned out, that was God's blessing for us. My wife has since given birth to our second son and they are doing very well. I was fortunate to be nominated and selected to represent my country in an international expedition to the Amazon Basin for natural rubber germplasm collection from January to April 1981. We had to quarantine for one week in a temperate country, and so while all my international counterparts went to Britain, I sought permission and got approved to spend my week in Canada. This is the temperate country where I made new friends over the past three years. I felt it would also give me an opportunity to thank those who were very helpful to me and my family during our stay in Guelph." The bank manager listened with an unusual attentiveness as I continued my conversation this early in the morning of a new banking day. Then I said: "Sir, on behalf of my wife and family and my very good friends, Brian and Kathy, I want to express my deep appreciation to you for taking a risk on us by granting us a loan just when no one would have trusted that we will be able to repay the loan from such a long distance. I like to say that with the special grace of God, I am here to pay-off the loan as well as visit my alma mater."

I could see the manager lighten up and seat up straight with a potential restiveness giving way to a smile recalling how long I had already engaged him in the middle of the customers in the bank that morning. I presented the travelers' checks to him. He went through them and said: "Mr. Onokpise, you have six thousand U.S. dollars worth of traveler's checks. Do you really want to cash all of these? According to our records (he had already pulled up my records during the time we had been talking) shows that you owe the bank only eight hundred Canadian dollars. Do you intend to deposit the rest to your account?' Michael, the Manager asked with a hesitant look in his eyes. "I will leave some in my account but I plan to use most of it to get something for my family. Since we returned to Nigeria, we have not had a car. So I plan to go to Toronto tomorrow morning in search of a used car, which I will then ship to Nigeria. I said earlier, this has been God's blessing." I replied. "That is very good, Mr. Onokpise. I know a used-car dealership in Toronto with whom our bank works in securing car loans for some of our Guelph students and staff. I will give you a note to him. I am sure he will give you a good deal. Once you have concluded your transactions you can come back to get the bank certified check." He concluded his statements and called one of the tellers to handle the travelers' checks and the loan pay-off. Then he gave me the note to the car dealer and asked "How are you are going to Toronto?" "My family friend, Gary will be taking me to Toronto, later this afternoon" I replied.

With the bank transaction completed and especially the loan paid-off I walked briskly to my former Department of Crop Science. There, I went up to the Departmental Chair, Dr. Kasha who informed me that my major professor had gone to Europe for a conference. He asked me all kinds of questions including the usual one as to whether I was going to obtain my Ph.D., now that I had already achieved international scientist status in such a short time. I told him that I did not have the same notoriety as my esteemed professors to whom I aspired to be someday. I informed him that I had been accepted to study for my Ph.D. in Tree Breeding and Forest Genetics at Iowa State University starting in the fall of 1981. Dr. Kasha was a highly regarded international Cytogeneticist. I was a student of his Cytogenetics course required for all M.S. and Ph.D. students in Plant Breeding and Genetics. After several exchanges, he wished me the very best in my up coming Ph.D. program and further successes in my professional career. Then I walked through the three floors of the building housing the department. I also went to the greenhouse to say thank you to the greenhouse manager for the practical help he gave me during my study in Guelph. I was told by one of the students working there that he had just left for a brief coffee break. I stood by the very bench where three years earlier, I had made twelve thousand pollinations transferring pollen from one flower to the other in order to produce offsprings of the forage legume plant I conducted my research on. Standing there brought back a lot of memories.

I looked at my watch and it was almost 12:00 noon. I immediately ran back to the fourth floor of the Crop Science Building, where the Environmental Science Program was also housed. I needed to see a Fillipino lady who was studying for her Ph.D. She and her husband had become our close family friends. I wanted to find out if she had finally resolved some of her problems related to passing the qualifying exams. The lab technician told me that she had not seen her for a few days. I asked her for Celina's (the Fillipino student's name) phone number and home address. She (the laboratory technician) was kind enough to give them to me after asking me several questions. I called Celina who picked up the phone after about four rings. "Celina, this is Kome, I am in Guelph for a few days and would like to visit you and your family later in the week." I said to her. In reply, she said "Kome, what are you doing in Canada? I thought you had returned to your job in Nigeria. Yes, it will be nice to see you again. A lot has happened since the last time we spoke before you and your wife left for Nigeria. "We then set up the following evening for my visit to her place.

Thereafter, I called Gary who had agreed to accompany me to Toronto to purchase my car. Gary and Cindy were a very young couple with whom we shared the same floor at the Westwood Apartments. Two other apartments separated our apartments from each other. Gary was an Ontario native from the farmlands in Kemptville, Ontario, while Cindy was from Montreal. Her beauty was striking which made us wonder how both of them met. Cindy's father was a Professor at McGill University in Montreal while her mother taught piano lessons at a private preparatory school also in Montreal. Their backgrounds couldn't be much more contrasting, yet they were one of the most loving young couples one could ever meet. Gary was in the same department as myself studying Plant Physiology. We took one or two courses together. In the 18 months we stayed at the Westwood apartment Cindy had become a very good friend of Lucy and always stopped by our apartment to help Lucy with babysitting our two year old son, Maro. So, that evening after I checked into my hotel while preparing to go for dinner I had called Gary who was home and had promised to take me to Toronto once I was ready to do so. We talked about old times and looked forward to driving together to Toronto.

Gary was in the growth room when I finally caught up with him at about 1:15 p.m. "Gary!!! Gary!!!!" I screamed. We had not seen each other since they accompanied us to the Toronto airport for departure to Lagos in 1980. He had some potted seedlings he was transplanting. "Kome, welcome back to Canada. It is nice to see you again. How is Lucy doing? She must have given birth to your second child. How is that your cute baby boy?" In return I asked about Cindy and how she was doing on her job and Gary with his studies. Gary and I hugged each other. "Let me finish these last ten pots and I will clean up quickly and we can start heading for Toronto." Gary said in his usual hoax voice. "You take five and I will help you with the other five. Just show me exactly what you are doing. After all, just 18 months ago I was transplanting and watering plants in the same growth room." I replied. Gary looked at me, smiled and transplanted the next seedling as I watched very closely. Thereafter, I worked with him and within ten minutes all the seedlings were transplanted. Then I jokingly told him that if there were an error in the outcome of this particular part of his research he would know exactly where it came from. "Kome, wash your hands and let's get going. Five seedlings out of two thousands will not make that much of a difference" he retorted. "By the way, have you had lunch?" Gary quickly asked. "No." I replied. "Then, let's stop at the university cafetaria for old times and grab a lunch before leaving for Toronto. I am starving. I have been in the growth room since 7:00 a.m., so other than a cup of coffee, I have not eaten anything." Gary responded. "That will be fine with me," I replied. So we went to the University Center for lunch. It was 2:00 p.m. when we drove out of campus. Toronto was only 63 kilometers (about 40 miles) from Guelph and within an hour we were there.

Using the note and address the bank manager gave me earlier in the day we drove straight to the car dealership. Upon arrival we received a warm welcome and the usual sales pitch for a car on a car dealer's lot. "How can I help you, gentlemen?" as John asked Gary and I went to sit on the chairs right across from him. Then I told him that Michael, the branch manager of the Canada Imperial Bank of Commerce on the campus of University of Guelph had referred me to him. I pulled out the note from Michael, the Bank manager, and handed it over to John. As he read the note I noticed that his eyes lightened up and smiled as he finished and folded the note.

"Okay, Mr. ------- I am sorry I cannot pronounce your name." Just then, Gary cut in "We call him Kome." "Thank you, Gary." John replied. "Alright, Kome, you are going to ship the car immediately you pay for it." John asked looking straight at me. "Yes." Was my nervous response, not having bought a car previously from a car dealer in Canada. "You are asking for a used Toyota. Those are very good cars. We have three or four used Corollas on the lot. Those cars often leave the lot as fast as we get them. We received these ones just last week and already two of them may be sold within the next few days." John concluded his sales pitch. Of course he didn't have to tell me about a Toyota vehicle. I owned a Toyota Corona for two and half years and sold it when I had to go for my M.S. degree. I loved the car very much and I got nearly 80 % of what I paid for it when I bought it brand new.

We followed him out of his office into the lot and walked towards the cars. After checking all four cars I chose the grey colored corolla. He opened it and told me all about it as to the previous owner and the number of miles on it. Overall the car was in very good condition. We went back into his office to complete all the paperwork for purchasing the car. This used 1978 Corrolla was offered to me for four thousand Canadian dollars, which was about three thousand and three hundred US dollars. I felt very good about the sales. John did not add Canadian taxes and registration charges since I was shipping the car out of Canada within two weeks of purchase. Furthermore, he said I received excellent discount because I was going to pay cash. He promised to get the bill of laden and all papers for exporting vehicles ready for me so that once I picked up the car from his lot, I could ship it. Meanwhile, he had already contacted a shipping company, which indicated that shipping to Nigeria would be taking place within seven days. He would confirm this by the time I got back the next day to take possession of "my new car." He gave me the papers I had to present to the bank manager, shook my hands while congratulating me on the purchase of the car and wished us a safe trip back to Guelph. With broad smiles, we all wished each other goodbye.

It was already 6:00 p.m. when Gary and I got back on highway 401 and headed for Guelph. Gary called Cindy to prepare dinner and so when we got to Guelph, dinner was ready. "Welcome back, guys!" Cindy said with a very broad smile as she opened the door and kissed Gary at the door. "Kome, how was your day? I hope you were successful in purchasing the car." She asked while directing us to the dining table. Gary and me were very hungry and we sat down to dinner even before answering the questions Cindy was asking us. After a glass of water and a sip of some orange juice, I took a deep breath and started to respond to Cindy's questions as she passed the food around to Gary and I with smiles that made her beauty even more stunning. "Yes, Cindy, we were very successful. We purchased a 1978 used Toyota Corolla. It is in very good condition. Thanks to your husband, Gary, and John, the vivacious salesman who worked hard to make our journey very rewarding." I responded as I started eating the first slice of freshly prepared French bread. "Lucy is going to like the car considering the 1968 Ford Maverick you all owned while you were living here in Canada." Cindy commented. All three of us laughed at that satire, knowing very well the condition of the Maverick. "Cindy, you know I was a graduate student and that's all I could afford but it served us very well." Again, we all laughed and settled into the dinner of fried rice, potatoes, and gravy. In keeping with her French tradition Cindy served wine with the dinner. We finished dinner and as I got ready to leave for my hotel, Cindy asked one more question. "By the way, how is Maro doing since returning to Nigeria? He must be growing fast? Has he started school now?" I answered yes to all the questions and said "Cindy, thank you very much for this very delicious dinner. Good night" as I waved to her and Gary closed the door behind us while we went to the parking lot for Gary to give me a ride back to my hotel. I waved good night to Gary as he dropped me off and thanked him immensely for helping me out in purchasing the car. We both agreed that I would take a taxi the next day back to Toronto to pick up my car so that he could finish with his research project in the greenhouse. Once I entered into my hotel room, I undressed, put on my pyjamas and went straight to sleep after such a long and rewarding day.

The ray of sunlight coming through the window woke me up the next morning. It was already 8:00 a.m. I knew I had to be in Toronto at about 2:00 p.m. This was to allow John to complete all the paperwork and finalize the date for shipping my car to Nigeria. Meanwhile I needed to be one of the first people to get to the bank on campus in order to get a certified check to take to Toronto. Without turning on the TV, I went into the bathroom after my usual morning prayers. Nearly four months in the Amazon jungle made morning TV not as interesting as it used to be. By 10:00 a.m. I was already at the bank. The lady at the counter motioned me to come up and asked how she could be of help. "Thank you very much ma'm. I need to speak with the manager." I told her. "Please wait, he is presently attending to a customer and he will be with you as soon as he has finished with her." She replied.

Within 10 minutes I saw a young lady come out of the manager's office. She looked at me as though she knew me from somewhere. I too gave her the look of familiarity. She took several steps forward and then stepped back. "Aren't you Kome from the Department of Crop Science? Didn't you already complete your M.Sc. and returned to Nigeria? What are you doing here? I am going to the cafetaria for brunch, can you join me there when you finish?" She concluded her questions. "Yes, Patricia, I will. I haven't had breakfast yet. It will be nice to talk with you again." Patricia and I took a Statistics course (Applied Experimental Design) together in the fall semester of 1978. Just as she completed her statement the bank teller called me and indicated that the bank manager was ready to see me.

"Kome, good morning. How was your trip to Toronto? Were you able to identify a car and how did John treat you?" Michael asked me these questions as he shook my hands and motioned me to a chair opposite him. I smiled broadly, pulled out all the paper work and replied. "It was a very fruitful trip. John was very professional and extremely helpful. His efforts coupled with Gary's presence to provide support, enabled me to find what I was looking for. The price was also reasonable. As you instructed, I have brought back all the paperwork needed to generate the certified check to enable me pay for and ship the car." I concluded my statements while handing the envelop with all the documents to Michael. After studying the papers he called on one of the tellers to prepare the certified check. Within 30 minutes I had a certified check in my hand. John had been very nice to call up the shipping company at the same time that he was also selling me the car. It would appear that he had concluded arrangements with the shipping company so that he would charge me for both car purchase and shipping costs. "Kome, here is the certified check. Congratulations for your car. I am pleased to have a customer like you. I wish you all the best in your future professional career. Remember, you are always welcome here." Michael commented as he handed me the check, which I had already signed for. He shook my hands and gently waved to me as I left his office with smiles, gratitude, and a sense of accomplishment with a great burden lifted.

Patricia started waving to me as I walked into the cafeteria and invited me to where she was sitting. I immediately signaled to her that I would be right at her table once I paid for my food. After selecting what I considered was a well-deserved brunch (pancakes, two susages, fried egg, a banana, a glass of orange juice, and water), I paid and went straight to share a table with Patricia. "Kome, I thought I would never see you again after our last encounter in the library. How is your family doing? I knew you finished and went back to Nigeria. We found out on graduation day when your name was called but you were not there because I personally did everything to see you walking on that stage. Later on, some body told me that you and your family had to leave as soon as you defended your thesis because of the commitments to your employers who sponsored you." Patricia kept on going until I had to interject with a response. "Yes, Patricia, my employers had already sent air tickets for my family and me and I had a time limit to be sponsored especially as the sponsorship also provided for my family. With such support and a job waiting for me coupled with the fact that my major professor worked very hard with me to get my research on track from the beginning made everything come together and so I had to leave as soon as I defended my thesis. It was a very difficult decision to miss the commencement exercise, but Guelph will always be a favorite place for me. I made a lot of very valuable friends including you, Patricia." I stopped talking as I felt my food was getting cold.

Patricia and I had taken a Statistics (Applied Experimental Design) course together just before my family joined me from Nigeria in 1978. For some inexplicable reason, we were always grouped together for laboratory exercises and other assignments. So we got used to talking to each other a lot of times. One afternoon during one of the activities for the African Students Association annual celebration on campus, I modeled a dress with my three-year old son never thinking that Patricia would have been out there watching the activities at the students' center. As I stepped down from the stage with my son, here was Patricia with those usual questions "Is he your son? Are you married? Is your wife here with you? How come you never told me you had a family? Any way you both look beautiful in those dresses." Patricia concluded. To say that I was a little shocked and embarrassed will be an understatement. At the time I never thought anything about Patricia beyond just being course mates who happened to have been taking a class together. I was so focused on finishing my program coupled with the fact that my wife and I who were in our third year of marriage and raising a son in a foreign land, that the thought of any other relationship was the farthest thing from my mind. I looked at her and answered yes to every question she asked. "Patricia. I had to protect my family and focus on my studies. I apologize if I did not let you know that I was here in Canada with my family. I hope this will not hurt our friendship, as working together has helped me in the statistics course." I told her and stretched my hand to shake her. "No---o---o, No—o!!! I just wondered. You are a very gentleman. Both of us studying together have been helpful to me too." Patricia responded and shook my hands in return.

Here we are a year later eating on the same table and still reflecting on the events of two years earlier. We laughed and joked about our short time together in class and how we had a great time in statistics. By now we had finished eating and getting ready to empty our trays into a trash can when from no where Patricia threw me a curve ball. "Where are you going to be this evening? I am staying with my sister for a day or two as I pick up my transcripts and visit some of my former professors to get letters of recommendation. I am applying to McGill University in Montreal for my Ph.D. program. May be we can rendezvous later." I stopped, held down my tray and took a straight look at Patricia who also gave me the smile as if to say "why are you so surprised?" I smiled back at her after collecting my thoughts and said "Patricia, I am glad to hear that you are applying to a very good Canadian University for your Ph.D. I hope your French is good. I am also starting my Ph.D. at Iowa State University in the Fall Semester. In fact, I deferred the start of my program by a year in order to participate in the important Brazil expedition which I have just spoken to you about." She was still staring at me, eyes wide open and smiling waiting for me to say yes or no to her rendezvous question. Then I continued my statement "I am going to Toronto to pick up a used car which I purchased yesterday and prepare it for shipping to Nigeria. I already have my evening full with appointments to my family friends, so it will be very difficult for me to rendezvous with you later today. However we could have lunch together tomorrow just before I leave for Montreal from where my car will be shipped." It would appear that Patricia already knew that it was not going to be possible for another meeting because she had a very busy two days ahead of her and this looked like the only time both of us were going to see each other. We stood up, took our trays and walked in steps to the trash can and emptied our trays. We looked at each other simultaneously, and smiled voluminously. We embraced each other for about two minutes and as we pulled away from our embrace, Patricia remarked "Kome, it has been nice knowing you, greet your family when you get back to Nigeria and I wish you all the best in your future endeavors." I replied with the same emotional voice. "Patricia, it has also been very nice to have met you. I know you will be successful in life and wish you the very best in all that you do in the future." We pulled away from each other, shook our hands and said goodbye as she walked away and I turned to use the paid phone to call a cab for my Toronto trip.

By the time the cab arrived, it was already 1:00 p.m. I gave the cab driver directions to the dealership in Toronto. We arrived at about 2:30 p.m. having been caught in the traffic due to some construction work just outside of the city limits. I paid the cab driver and went straight to John's office. He had just returned from lunch break, so he was attending to no other customers. It was perfect for me as his attention was focused on finalizing my car purchase and preparing the shipping documents. John had every thing under control. Once I handed him the certified check, he smiled and said, "This is very good. I like working with customers like you. By the way, your car is ready to be driven out of the lot. Are you ready?" My immediate response was "Yes!!" Then I asked him about the promise he made to ensure that my car got on the boat for Nigeria. "No problem, I have your bill of laden here. The company has an office here in Toronto and we work with them all the time to ship new and used cars to all parts of the world." You need to have the car to them by mid-day Saturday so that it can be loaded on the ship. Here are all your documents for the car and the bill of laden for shipping." John concluded. With keys in his hands, we stepped into the lot and in less than three minutes, he stretched out his hands and said. "Here is your car. Congratulations!! Remember to drop us a line once you clear it from the port in Lagos and you are enjoying another Toyota." I knew he had to put in a last minute advert for Toyota. I smiled, took the keys from him, opened the driver's door and started the car, having already driven it around the previous day. I stepped back out and said, while shaking his hands "John, thank you very much for all you have done. I am not sure if I could have found a better deal in Toronto. Furthermore, helping me to arrange the shipping of the car is a great relief given the fact that I had only one week to get all of these things done. Thank you very much. I will definitely drop you a line as to how well this sale went and how good the car performs on Nigerian roads. I have no doubt it will be okay." I concluded my statement.

I got into the car and waved to him as I drove off the lot. I still had my Canadian driver's license on me so there was no problem with driving in Canada. I was also familiar with a few streets in Toronto having come to downtown Toronto about once a month to the open farmers market to purchase traditional and speciality food items for my family during my two years in Canada. I did a little bit of shopping, picking up souvenirs and items for our new born baby and our four-year old son. It was nearly 5:00 p.m. when I arrived in Guelph. I called Celina from a paid phone to inform her that I was going to visit briefly with her and her husband that evening since I had to see my other family friend that same night. She responded in the affirmative. Her voice did not seem excited which made me wonder if something was wrong. I found a parking place in the hotel not too far from my room. I went to the lobby to get the keys to my room and the middle aged lady at the front desk took a look at me and said "Son, I noticed you just drove a car into the hotel parking area. Did you just buy one?" "Yes ma." I replied. "So you can now find you a nice young lady for the rest of your stay. You have been here for four nights now and I haven't seen you with any lady. A young man like you should be with some loving woman to keep you warm at night." She responded with a very broad smile. I thought to myself, Patricia had asked for a rendezvous this evening, but I told her I had already made commitments so could not take her up on her offer. This woman was reflecting the same sentiment. After a moment of silence I responded to her "Ma, I am a married man and have a family so I am not going to be having a woman or lady even with my new used car. I appreciate your concern for me. I am going to my room to shower and then go visit my family friends before coming back to prepare for my journey to Montreal tomorrow." "Okay, what can I say?" She finally responded.

At about 6:00 p.m. I was back in my car heading for Celina's house. I rang the doorbell and waited for about three minutes before knocking on the door. Celina opened the door and motioned for me to enter into the living room. She looked down as she asked me to take a seat and in a very shaky voice said "Kome, it has been nearly 10 months since you and your family left for Nigeria. How are they doing? What brings you to Canada? Are you planning to return for your Ph.D.?" "Celina, first I like to apologize for not coming to visit you two nights ago as I had promised. With no vehicle, I could not go anywhere once I returned to my hotel at the end of the day. In any case I am here and wanted to say hello to you and your husband. By the way where is Manuel? How did your Ph.D. studies go?" I suddenly stopped asking further questions immediately I noticed that Celina was crying. Then she got a piece of towel and wiped tears from her eyes and face. Quietly, she started to speak. "Kome, a lot has happened since you left. As you know, just before you left I had taken my qualifying exam for the Ph.D. and there were some issues. I tried to change my major professor with little success. In the end I had to give up on pursuing my Ph.D. Fortunately, I found a nice paying job. Manuel also found a very nice job at a factory near Toronto. We were doing quite well as a family with our children starting High School. We turned our attention to begin to help our extended family in the Phillipines.

Then, one afternoon I got a call from the factory where Manuel worked. The news was not good. Equipment had accidentally slashed through Manuel's right arm and before the emergency vehicle arrived, Manuel had bled so much that by the time they got to the hospital it was too late. He did not make it ---------" Celina could not finish the sentence as she choked and cried, this time with more intensity. I realized that at this point a lot was now going through her mind, which is why she was crying more intensely. I picked up the towel for her and consoled her. She paused, cleaned her face again and was about to resume her talk when I said. "Celina, that is fine. My condolenses. It is sad what has happened to you in a very short time. Our prayers are with you and May God continue to bless and be with you." Once she had recovered emotionally, I explained to her why I was in Canada and all that had transpired within the four days I had spent in Guelph. Then I said to her. "I will be leaving for Montreal tomorrow in order to catch my flight for Frankfurt en route to Lagos after placing my car in the boat for shipping to Nigeria. I have also concluded arrangements to continue studies for a Ph.D. at Iowa State University, in the United States through a scholarship from the Federal Government of Nigeria. I had to defer the start of my program to the fall of 1981 in order to undertake this important expedition. Celina, I saw a lot in the Amazon Basin. It is interesting to see what happens in the remotest parts of the world. It was also a very rewarding experience for me." I noticed that Celina was beginning to lighten up as I recounted my experiences of the germplasm collection to her.

I supposed it brought back memories to her as an accomplished biologist who still contemplated returning to graduate school to complete her Ph.D. in Environmental Biology. Then we talked about other things and laughed as to how things work out in environments unknown to us. Just when she remembered that she had not offered me a drink, I checked my watch and realized that I needed to be on my way to catch a dinner appointment with Gary and Cindy. There was also forecast for snow that evening. "Celina, I have an 8:00 p.m. appointment with Gary and his wife and I am already running pretty close to the time. It has been very nice to speak with you. Please keep up your spirits and stay strong for your children. Do not forget to attend mass as you and your family, have always done. We always continue to pray for God's assistance." I concluded my statement. "Kome, I didn't even offer you a drink. You don't know how much your visit uplifted my spirit even though I was emotional in the beginning. I hope you didn't mind that. I wish you all the best in all your future endeavors." By now we were already at the door and as she opened it she said, "Thank you very much and please kindly remember to greet your wife for me. I really miss her. Drive safely to Montreal and fly safe to Nigeria." "Thank you very much and I will tell Lucy you extend your greetings to her." I replied. We waved goodbye to each other as she closed the door and I walked to my car.

By the time I arrived at Gary's, Cindy had already served food on the dining table. "Kome, welcome. How was your day? I can see you are all smiles. How is the car driving?" Gary asked in his usual jovial way while we took our seats at the dining table. "Gary, I picked up the car and John was fantastic. He got me all the details for managing the car as well as the bill of laden for shipping the car and eventually clearing customs to obtain the car in Lagos. Thank you very much for everything. No doubt your presence yesterday went a long way to ensure that I was well taken care of. By the way, how is your research going? I hope you were able to complete the planting of your experimental trials. Please send me a reprint of the paper when you eventually publish your paper." I replied. Immediately we finished dinner, Cindy who had been unusually quiet during the dinner said smiling "Guys, I think I should say a word or two. Kome, it has been very nice to see you here again. Remember to tell my friend, Lucy that I really miss her and Maro. I hope when you come to the U.S. for your Ph.D. program we can visit each other and continue our friendship. Gary has already been offered a job at the Agricultural Research Station in Kemptville. So we will be in Ontario for a while. Again, thank you for coming." "Cindy, thank you very much for your kind words and I will definitely relay your message to Lucy. I am extremely grateful for the warm reception and hospitality you and Gary have given me during my brief visit here. I have great memories of Guelph and Canada and so any opportunity my family and I have to come back here we will." I replied. Then turning to Gary, I said "Congratulations!!! You didn't tell me about your job offer. It is always nice to have a job lined up before one graduates. I am sure you are going to be a great research scientist." Gary, stroking his long beard in his unique way, laughed out loud and said humorously "You were so focused on getting that Toyota, that anything said to you may have been forgotten anyway. Thanks for your congratulating words." Then I reached into my bag and pulled out two University of Guelph t-shirts I had bought earlier in the day and presented them to both Gary and Cindy. These were very well received as both Gary and Cindy thanked me simultaneously and indicated that they had been planning to get some t-shirts from the university but had not got around to doing so. Therefore my gift was very timely.

The clock on the wall struck 10:00 p.m. This reminded all three of us that we had places and things to do the next morning. Cindy had already cleared the dinning table even as she spoke to us. She came to the living room and we all hugged, as we got emotional knowing fully well we were ready to wish each other goodbye. Cindy said "Kome, I know you are going to be successful, just keep us in mind and let us continue this friendship. You are going to spend a full day in Montreal. It is a very beautiful city. You should not have much problem since you speak French quite fluently. Compared to Quebec City, French language in Montreal is much easier to understand. I suggest you drive around the city a little bit before you take your car to the sea port for shipping. Bonne nuit et Bon Voyage!!!" Cindy concluded her comments as she walked to the bedroom to sleep. I also said "Bonne nuit" and smiled back while thanking her for her patience and the very tasty French style dinner she exposed me to. I promised her that I would take full advantage of the opportunity to see as much of Montreal as I could. Meanwhile, Gary accompanied me to the parking lot and once I got into my car we shook hands and wished each other well in our chosen professions. I told him I would continue to maintain communication. I even told him that he should let Brian know that I visited and I hope to see him someday in the U.S. Gary and I waved goodbye. I drove back to my hotel. It was already 11:00 pm. when I got into my room. I decided to go to bed right away considering I had a long drive ahead of me the next day.

It is now five days since I arrived from Manaus to Guelph. I had basically accomplished all the things I wanted and why I chose to come to Canada for the quarantine. I went downstairs to the hotel restaurant to get one of those free continental breakfasts. Following breakfast, I went to my room and packed up the luggage lying all over the room. I realized I had accumulated a lot of materials in a very short time, having bought presents for my family both nuclear and extended. Had I not bought the car, I would have ended up paying a lot for excess luggage. I loaded as much materials that may not be needed immediately when I arrive in Benin City, into the car and prepared them for shipping. I went to the check out counter and found the same receptionist who wondered why I had not a warm body with me throughout my stay in the hotel. "Hello young man. Are you checking out? Your wife must be a very lucky woman considering how disciplined you have been this past five days even after purchasing your car. In any case it has been very nice to have you here. We will like to have you back. Always remember us in Guelph. By the way I like your blue and white University of Guelph jacket." She concluded her comments. "Thank you very much for your kind words. Yes, I will extend your greetings to my wife and children. Guelph has a special place in my life and I will make every effort to return to Guelph whenever I get the opportunity to do so. I appreciate everything you all have done. I enjoyed my stay here." I responded and collected my receipts from her. I waved to her and got into my car for a drive to Toronto where I had planned to spend the night and leave early the next day for Montreal.

Once I drove onto highway 401 I realized I was going to need some music to keep me engaged while driving for nearly five hours. I remembered that I had seen two eight tracks one reading ABBA and the other AC/DC in the globe compartment of the car. ABBA I was very familiar with. In fact I had the cassettes and albums of ABBA but never AC/DC. Naturally, I slotted the ABBA eight track first and enjoyed the music all the way to Toronto that afternoon. Because I had already reserved my hotel room just on the outskirts of Toronto, so I could get on 401 quickly I had no traffic delays, arriving in my hotel within 45 minutes of leaving Guelph. It was 6:00 p.m. I made sure my car was well locked looking at all the items I had purchased. I settled in with little or no funfair. I pulled out the note pad I had with me and made some notes and revisited what I had written down throughout the nearly three months in the Amazon basin. I made some notes, rechecked my air ticket schedule and went to the hotel restaurant for dinner. There after I returned to my hotel room, set the alarm for 5:00 a.m. Instead of watching another TV program I went to bed knowing fully well that I had nearly five hours of drive the next day. As soon as the alarm clock rang I woke up, said my usual morning prayers and took my showers. I was on highway 401 headed for Montreal at 6:00 a.m. I had already listened to both the early morning BBC and CBC news on radio so after briefly stopping for breakfast, I slotted the AC/DC eight tracks into the car system. Initially the music sounded too loud and a little out of place for me. Yet, with every mile I drove I started to feel comfortable with the sound and over the next 30 minutes I got used to the rhythm of the music. This was my introduction to heavy metal music.

I arrived in Montreal at 1:00 p.m. on April 25, 1981. My hotel check-in was 3:00 p.m. so I took a power nap inside the hotel packing lot just to be on the safe side given the fact that I did not know much of Montreal. Once I woke up, I remembered Cindy telling me to see as much of Montreal as possible. Because I had sometime left to check-in I got out my compass that I had carried with me from Brazil, set the location of the hotel and then went on a two hour drive around the city. The bridges were the most striking for me followed by the magnificent cathedrals. It was 5:00 p.m. by the time I got back to the hotel. I walked up to the counter and as I expected, the young lady at the desk said "Bonjour monsieur. Soyez le bienvenu au Montreal. Avez vous reservez une chambre a notre hotel?" I replied in French simply "Oui!" then I quickly asked "Parlez vous anglais?" When she responded "Yes" in English I felt quite comfortable. It reminded me of Cindy's words that I would find more English speaking Quebecois in Montreal than in Quebec City or the other parts of Quebec Province. I was checked into my hotel room. I went into my hotel room with only the two luggages with which I was going to take my flight back to Nigeria.

After dinner that evening I decided to watch some TV programs. Then I stumbled onto this channel called Cable News Network or CNN that was running twenty fours news cycle. I recalled that in the two years we had been in Guelph there was no such network. What caught my eyes was the report of the first space shuttle, Columbia which had been launched into outer space just a few weeks earlier before my arrival in Canada for Quarantine. I had always loved discussions on Space Exploration since my secondary school days when I led a Literary and Debating Team to win an interschool debate titled "Is Space Exploration Justifiable?" in 1968. The following year, 1969, Neil Armstrong and his fellow Astronauts landed on the moon in Apollo 11. It was a great excitement for us. So, here I was in Montreal, quarantining and enjoying a space exploration story. I watched the entire show until I fell asleep. I was woken up the next morning by the bright rays of the early morning sun coming through the hotel window. It was already 9:00 a.m. I reminded myself that the car needed to be at the seaport at 12:00 noon in order to be loaded onto the ship for shipment to Lagos. I got the car to the port just in time. They checked all the paper work and I was to drive the car onto the left deck of the ship. By placing the car on the deck by myself, I felt confident that the car would be shipped. At the hotel, I had requested a late checkout so that I could take an evening cab to the international airport in Montreal.

I took my seat in the economy class of Lufthansa, an airline my family and I had flown several times and we had come to love it. As the plane took off I thought about how fast the quarantine week had gone and here I was returning to Nigeria having spent three months in the remotest parts of the Amazon Basin, and an exciting and fast moving week in Canada. With these thoughts going through my mind I fell asleep and only woke up when the air hostesses announced "Prepare for landing at the Frankfurt International Airport." Shortly thereafter the plane touched down. It was about 10:00 a.m. local time when we arrived in Frankfurt. Because my connecting flight was not until 2:30 p.m., I had four hours lay over time. I remained at the airport and experienced the enormity of the Frankfurt International airport.

# CHAPTER TWELVE

## RETURN TO NIGERIA

Lufthansa touched down at the Lagos Muritala Mohammed International Airport at about 6:00 p.m. on April 27, 1981. As I stepped out of the plane into the tamarc to get on the bus that took us to the waiting room for our luggage, I realized that the great "Germplasm 81" had come to an end. I was now back in Nigeria and needed to prepare to meet my family. Through the technology of telegraphic messages I had already given my flight itinerary and arrival date to my sister, Mrs. Joe Akiti and her husband in Lagos, and my wife, Lucy, in Benin City. So, I anticipated that my sister and her husband would be at the airport. They were not the only ones there. My mother was also with them. Thus, when I came out of the luggage area it became a very rewarding homecoming from one of my most important international assignments. Interestingly, it was my mother who was the first to greet and ask me questions. Embracing me very warmly, smiling and patting me on the back with a sense of pride she said, "My son, welcome back home. In one trip you have covered four continents, so they tell me. This must be a wonderful way to work and earn a living. You are now like your late father who used to travel up and down the Niger and Benue Rivers. Your sister and her husband were talking about you yesterday as to how in just four years you have already been to more countries as many people will visit in their lifetime. We must thank God for all His blessings." "Mama, I thank God for all the opportunities He is blessing me and my family with. You know you and papa raised a wonderful son" I jokingly concluded my statement. By now we had reached the car my brother in-law and my sister drove to the airport to pick me up. Ernest, my sister's husband welcomed me formally so did my sister. My mother was allowed to get into the car first. I sat with her on the back seat. All the way back to their apartment (flat) I told them what happened throughout my three months including the one week quarantine in Canada and how that helped me to completely release all the pressure of spending three months in the Amazon Basin.

My sister in her usual way said "Why are you so thin, you have lost so much weight. Didn't they feed you all in the jungle? Or was it hard work? Any way, once you get to Benin City make sure Lucy cooks good food for you so you can gain some weight." I replied. "Sister Joe, the scientific work we were engaged in was very intensive. We worked an average of 12 hours a day from dawn to dusk covering 20-25 kilometers (12-15 miles) each day carrying nearly 70 kilograms of materials in our backpacks. As you can see that was a lot of work. But I enjoyed it, walking through the thickest forests in the world. I know I lost close to four kilograms (10 pounds). Hopefully, once I resume my work in Benin City, I will regain some of my weight back." My mother added with a broad smile "Joe, you know your brother has always been thin so just losing a few pounds can make him look thinner than normal." We laughed through all the teasing and arrived at the apartment in FESTAC Village. The maid came to open the door and greeted us profusely.

My sister already had dinner prepared. So once we entered and settled down she served the dinner. Ernest was taking my luggage to the inner room. I reminded him that my flight to Benin City was scheduled for 10:00 a.m. the next day and therefore it was not necessary to store the luggage too far away. So he left both luggages on the far side of the dinning room. We had dinner and talked about all that has transpired in Nigeria while I was away. As dinner progressed, I broke the news that I bought a car during my quarantine in Canada. My sister with eyes wide open and a full mouth with food could not help herself and said. "Very good!!! I know how tough it has been for you and your family since you returned from your M.S. degree program. What kind of car did you buy this time?

Congratulations!! I see you had a very good time this past four months even though you look thinner." By this time she had emptied her mouth and was smiling and making a lot of jokes about the Brazilian expedition. "Joe, thank you very much for all your comments. My estacode (per diem) left over was sufficient to buy a used car and it is a Toyota Corolla. The important thing is that the car is being shipped from Montreal, Canada to Apapa, Lagos. I did the shipping that way instead of sending it through Warri or Sapele ports. I figured that since your husband is familiar with the customs officers and know how items are cleared from the ports given his job experience. It will be much easier for us to clear the car from the Apapa port faster once the car arrives." Both Ernest and my mother echoed my sister's congratulations. Ernest went on to say that it was a very good decision to ship the car to Apapa port and that in any case most imports into Nigeria came through this port.

It was already 11:30 p.m. and as we were about to retire and go to bed, it occurred to me that my mother should actually have been in Benin City with Lucy and all the children (Joe and mine). So I asked both Joe and Ernest "I have enjoyed mama welcoming me from Brazil. It is always nice to see mama in a joyful mood. But I thought she should have been in Benin City. What happened?" My sister responded "Ernest was on one of his Bendel State sales trip and as you know your place is where he normally stays when he is in Bendel State. He saw that Lucy had a four days off-duty. At the same time her sister, Clara, was also spending the mid term break with her. Therefore to give mama some break too, he decided to bring her to Lagos for a few days. The plan was that since you were returning this same week both of you will then fly back to Benin City. Kome, you know the last time mama flew was when you put her on her first plane flight from Benin City to Lagos on your way to Brazil. So she will be flying with you back to Benin City to complete the cycle you started." I was very excited to hear my sister's explanation that I simply said "Great!!" This was particularly so because I was making every effort to see that my mother got rid of her fear of flying. We went to bed at about midnight with the knowledge that we had to be at the domestic airport by 8:30 a.m. It was even more critical that we got to the airport well ahead of time considering I was now flying with our mother.

After checking in, and wishing my sister and her husband farewell, my mother and I sat at one end of the waiting room and reminisced over many things from my childhood to when we lost my father to the ghastly car accident to my evolution into an international scientist, something my late father would have been very proud of. Our flight was called up for boarding at 9:45 a.m. My mother and I sat side by side because I wanted to stay close to her. "Welcome to Nigerian Airways Flight number 555. The doors of the plane are closed and the pilot has asked us to prepare the plane for take-off. Please make sure your seat belts are fastened and all your seats are brought up to the upright position. We took off in less than five minutes" came the hostess's voice. My mother asked me a few more questions as I helped to fasten her seat belt. Just as I completed fastening the belt, the roar of the plane taking off could be heard from inside the plane. As with her first flight, I saw her holding very tight to her seat and grimacing. We lived very close to the airport in Benin City and always heard the planes taking-off and landing. Often, my mother would ask, "all that noise, how do the occupants inside the plane deal with the heavy noise if it can be heard this far and sometimes vibrating the house?" Having traveled by air for nearly two years, I would let her know that the passengers in the plane were just fine and that the noise had no impact on them.

Lagos to Benin City on the type of jet plane we flew in was usually a thirty minutes flight and by 10:30 a.m. the pilot announced that we were 15 minutes away from the Benin City airport and that the hostesses should prepare the plane for landing. Once again, the hostess announced that we should all fasten our belts and that they were coming around to collect all the trash and unused items. I quickly drank what was left of my orange juice and fastened my seat belt. I noticed that my mother had not eaten her snacks nor drank her juice. So I put them away and fastened her seat belt. No sooner did I complete that task when the hostess announced that we should be on the ground in five minutes. The word "ground" made my mother to suddenly turn to me with her eyes wide open as if to say what does she mean will be on the ground in five minutes. I let her know that we were close to Benin City and that we would be landing soon. Again the one to two minutes of the plane landing and taxiing on the tarmac to its parking spot were just as interesting for my mother as the initial take-off in Lagos. Once the plane came to a stop, I told my mother that we had arrived and that she could relax.

We collected our luggage from the baggage claim area and started moving towards the area where the cabs were packed when I heard "Kome!! Kome!! Kome!!!" It was my wife, Lucy. Although I had sent a telegram from Lagos to her about the flight and possible arrival time and with no phones to communicate at the time, I did not expect her to be at the airport especially with all the children she had to look after and the scheduled arrival time. Furthermore we had no car of our own since we returned from Canada almost a year earlier. Thus, I was pleasantly surprised. I quickly turned around and ran back to embrace her, excited at her coming to welcome my mother and me at the airport. She had our eight months old son with her. "Lucy, this is very surprising. How were you able to get ready and come to the airport so early in the day to welcome us? How is Maro and my sister's children? I hear Clara has been spending a short holiday with you. How is she doing? I am sure your mother has been stopping by to check on everyone." I kept talking and asking Lucy the usual homely questions until she reminded me that if we did not catch the next cab we may not get another one for the next hour and our 45 minutes flight could soon become a three-hour flight. By now, my mother was carrying her grandson and she watched us as we got the cab and loaded our luggage. Our flat located on Aghimien Avenue just opposite the famous Joromi Hotel off Ekenwan Road was only about two miles from the Benin City airport. So within ten minutes we were home. The cab driver unloaded our luggage. Lucy paid and said goodbye while thanking him. Meanwhile, my senior son, Maro, already ran out into the veranda and jumped on me. "Welcome papa" with a very broad smile. "Where have you been all these months? Mama told me that you went to a far country called Br—a—zi-l. Is that true?" As with a five year old he kept asking questions until the mother reminded him that his food was on the table and that he needed to finish eating, then come back and welcome his father again. I took him to the dinning table and sat him down to his meal. Then I greeted Clara, my sister in-law thanking her immensely for assisting my wife in taking care of her nephews and my sister's children while my mother was in Lagos. Clara was very happy to see me and greeted me profusely in the traditional Isoko language.

Later that evening once my mother and I settled down and were fully rested I described my entire journey and three months stay in Brazil especially how I managed to survive in the Amazon jungle. At one point she laughed and teasing said. "With all the forests and swamps in our native lands that would have been interesting reading about you lost in a jungle so far away." My mother didn't like that joke so my wife had to stop that and switched to talking about our friends in Canada. Her face lightened up when I told her I had dinner at Gary and Cindy's apartment. "How are they doing? Has Gary finished his program? Is Cindy still staying beautiful and talking about Montreal or Quebec? Did she play any piano music for you?" As Lucy continued her questions, I thought the time was right for me to give her the surprise news among all the things I had purchased including the few items which were in my luggage and which I had already distributed to her and our children. So,

I quickly went into the bedroom, pulled out the envelop containing the documents for our new car. Before she could say another word I gave her the envelop and said "Lucy, open the envelop and read the contents and tell me what you think." My mother looked intently as Lucy read the document. I waited to see Lucy's reaction. Then, just as I was beginning to respond to the questions about Gary and Cindy, Lucy jumped out of her seat, gave me a big hug and exclaimed. "Kome!!! Did you buy a car? Where did you get the money to do so!! Thank God!!! We can now go from place to place without standing in the sun or rain waiting for a taxi-cab." She danced back to her chair since she was sitting across from me because Maro had taken the space next to me. My mother clapped her hands and sang some Isoko songs and congratulated me again. Thereafter I told her all about Gary and Cindy and how Gary helped me in traveling to Toronto to help with the purchasing of the car. Finally, I said "all our friends in Canada whom I saw asked me to greet you and Maro. By the way I also paid off that loan which we took from the bank just before we left Canada. So, as you can see, this trip beyond being professionally successful it was also very good for our family. I am extremely grateful to God for His mercies. Let me also thank you, Lucy, for taking good care of not only our own two sons but also my sister's children." I concluded my statements. I turned to my mother and also thanked her as well for providing the motherly and grandmotherly support for my family. Dinner was served and we went to bed knowing fully well that in the midst of all the welcome I had to report to work the next day because I did not take any annual leave to coincide with my return from the Brazilian trip.

I woke up very early the next morning, got ready and walked to Mr. Ike Ugwa, my colleague's flat on Agho street which was the next street from Aghimien Avenue, to get a ride from him. He was just getting into his car when I got there. "Kome, when did you return? How was your trip? Please get into the car." He said as he opened the passenger side of his car for me. We talked all the way to the institute. The Rubber Research Institute of Nigeria (RRIN) is located in Iyanomo, a small village at mile 16 on the Benin City –Sapele road going south from Benin City. It was about 20-25 miles from our flats almost in the center of Benin City. By the time Ike and I got to the institute at 7:30 a.m. we had exhausted "all questions" to be asked on both sides. He asked me questions about the Brazilian trip while I asked him questions about the institute and what research had transpired during my absence. We both laughed and it was a very joyous morning as we stepped out and went into our offices to wait for the rest of the staff to come to work. Ike and I had become very close friends because of the unique pieces of advice he had given me on many occasions as the institutes Directors (Dr. Ayeke and Dr. Okaisabor respectfully) gave me increased responsibilities. Although two years my junior in terms of civil service seniority and a pupil research scientist in Soil Science yet he had taken time to work with me in developing research projects and observed how close my relationship was with our former and present directors given the level of assignments they often gave to me. In fact, three years earlier when I left for Canada to start my M.S. degree, it was Ike whom I asked to work with my wife and the institute to ensure that my family was able to join me in Canada within four months of my departure. Thus, we had become confidants for each other at the institute.

Immediately I got word that the director, Dr. Okaisabor had arrived I went to his personal secretary to make an appointment to see him in order to make an informal report of my Brazilian trip. She went into his office and came out smiling. "Although he had scheduled some meetings for the next hour, he said you should come in and give a brief report of your trip. He is very excited. He has asked me to tell the other people to wait when they come for the scheduled meeting. You have become a very important scientist around here." Stella concluded her statement and smiled again as she opened the door for me to go into the director's office. Standing up to my surprise, Dr. Okaisabor said. "Mr. Onokpise, welcome back from your unique Brazilian trip. How was the expedition? You look very good even though a little thinner. You know the work you and your fellow international rubber tree breeders just completed will have significant impact for the natural rubber industry for a very, very long time. Please sit down and let me hear about some of your experiences." Dr. Okaisabor shook my hands and pointed me to the chair right across from his table. With the level of excitement I saw on his face I wondered how I could describe "Germplasm 81" to fully satisfy my director who had skipped over his own friend and my senior by civil service classification, to select me upon the advise of Dr. Mekako, our senior plant breeder. Over the next ten minutes I described in detail all the activities from my arrival in Brasilia to the tremendous experiences in Rio Juruena and Rio Guapore, our reunion in Manaus and of course the quarantine in Canada. Finally, I thanked him immensely by saying "Sir, I am extremely grateful to you for having selected me to go to Brazil. I was able to work with the world's leading natural rubber research scientists and to visit some of the remotest parts of the Amazon basin. This for me is a life-time experience for an African Scientist as my counterparts often referred to me during this past three months. I will forever be indebted to you and RRIN. Thank you very much sir." Then Dr. Okaisabor requested me to prepare to give a one-hour seminar to the entire institute's scientists and staff. He called Stella to come back into his office. "Stella, please send out an announcement to the entire institute that Mr. Onokpise will give a seminar next week Tuesday on his Brazilian expedition and everyone should attend." Dr. Okaisabor instructed Stella. As she opened the door to go back to her desk, she reminded Dr. Okaisabor that his guests had arrived from Benin City. It turned out the guest was the chairman of the Bendel State Rubber Board. So without hesitation Dr. Okaisabor asked Stella to let him into his office even with me still in the office. Little did I know that the Director saw an opportunity to score some political points for our institute. Dr. Okaisabor stepped out of his chair and met Chief Igiebor at the door. "Good morning Chief. Welcome to Iyanomo. It is very nice to see you again. I apologize for keeping you waiting." Dr. Okaisabor said to the chief and turning to me, he said. "Mr. Onokpise, our young research scientist or may be not so young, just returned from Brazil as part of an international group of natural rubber research scientists who went to collect germplasm in order to broaden the genetic base of the existing rubber clones to avoid a potential crisis which may face the industry in the future." Chief Igiebor stretched out his hands and shook my hands with a very broad smile. I was already on my feet once my director left his table to welcome the chief. I shook the chief while bowing not only as a matter of traditional respect but also in honor of his position as the chair of the state rubber board which exerts tremendous influence in the affairs of the state dealing with such an important commodity crop for the economy of the state and the country. Once our formal introductions were completed and the chief was directed to the most comfortable seat, Dr. Okaisabor asked me to schedule another appointment with Stella because he wanted to discuss the trip further and he had other assignments for me. I thanked him again and bowing to the chief now sitting on the near end of the office I stepped out of the office and closed the door behind me. I spoke with Stella about the appointment and she kept staring at me as if to say, "you are now the most respected scientist at RRIN to be asked to come and talk with the director any time you want." We both smiled at each other, I thanked her for the nice comments and left for my office.

Dr. Mekako was already in my office waiting for me when I got back. "Good morning Sir!!!" I screamed at Dr. Mekako and shook his hands and genuflected in the traditional

Urhobo language greetings for a senior. "How are you doing? I heard you traveled to Lagos. That is why I did not stop by your office this morning. We have plenty to talk about. It was a great experience." I concluded my statement. "I came back last night and just got into my office when Stella came to call me saying that the director wanted to see me immediately. On my way back I will stop to chat with you so that you can tell me everything you did in Brazil and beyond." Dr. Mekako responded as he walked briskly to Dr. Okaisabor's office. That afternoon following his meeting with the director, Dr. Mekao and I spent well over two hours together recanting the entire germplasm collection exercise. I concluded by saying that "This is an experience which will stay with me for the rest of my life." Then he reminded me of the conversations we had during my selection and preparation for the trip. Finally, he said " as the most respected scientist in this institute, this is another feather on your cap. Once you obtain your Ph.D. in a few years and return to the institute the sky will be the limit for you. You have made all of us, especially those from the Delta part of the state, very proud. The director just told me that he wants you to intensify your research efforts particularly the studies across several locations (genotype x environment interaction of the first batch of clones from the earlier selections) which you started just before you went to Canada. Welcome back again. I look forward to the official seminar which Dr. Okaisabor told me he has asked you to present next week." That day went so fast that by the time Ike came to my office to give me a ride home for the day it felt like I had just begun my work for the day. In any case I returned home after my first day back on the job exhausted and feeling very satisfied.

Over the next several weeks, I gave my seminar, completed the report of my "Germplasm 81" and sent it to the IRRDB in London, England after being reviewed by Dr. Mekako and Dr. Okaisabor respectively. I started a germplasm nursery with some of the extra seeds I stored away in the deep reaches of my sport coat with which I traveled to Brazil. These included seeds from Juruena, Villa Bella and Aracatuba. In addition I had few seeds of other Hevea species such as *Hevea benthamiana* and *Hevea spruceana.* My agricultural superintendent and assistants were all very happy to see this new nursery started. Meanwhile, my travel to Akwete in Imo State where the RRIN sub-station was located and Etche in River State where we had planted one of the test plots intensified in response to Dr. Okaisabor's directives. In one instance we even had to transport budded rubber stumps in the official Range Rover which the director used for important guests, when the designated truck (or lorry) for doing so was deemed not fit to travel from Iyanomo to Akwete and Etche. It was a very exciting and busy time for me even as my wife and my family prepared to take advantage of the scholarship, fellowship and research assistantship awards which I had received to travel to the United States for my Ph.D. in Tree Breeding and Forest Genetics later in the year. The Brazilian expedition strengthened my resolve to succeed so much so that the family pressure of not returning overseas for an advanced degree were starting to wear on me. It was at that point I happened one afternoon to meet a cousin of mine, Mr. Wilson Agha, an accountant in training with the esteemed accountancy group known as Akintola Williams and Company. After several hours of discussion he reminded me that I was the new African Scientist who had been in the Amazon Basin and that the only way to solidify this accomplishment in the future is to obtain my Ph.D. He made it clear to me that these federal scholarships and state fellowships might not be available in the future if the political situation in the country changed. Our discussion also revealed to me that it was unlikely that my director would grant me another study leave with or without pay to travel overseas in just twelve months after I had just returned from Canada regardless of the new respect he had for me.

When I returned home that night, I called my wife and explained to her the situation facing us as a young family with tremendous responsibilities to our extended families. I spoke with my mother and with tears in her eyes she asked me to arrange a visit to my sister in Lagos to discuss this critical situation which will also impact her life while I am studying overseas for a second time. Suddenly a month after returning from Brazil my family's attention and mine had shifted to dealing with the implications of another major decision to be made which will not only affect my job but also the life style of mother, my siblings, my sister's children and most importantly my own family. Once I had the opportunity, I discussed the situation with Dr. Mekako, who had now become my mentor, and Dr. Okaisabor my director. Obviously, my cousin was right. No sympathy was shown. In fact, Dr. Okaisabor even felt I had not fulfilled my commitment to RRIN on the basis of the sponsorship I received for obtaining my M.S. degree. A few of my colleagues who had started to feel I was getting preferential treatment, stirred up issues as unfairness and my disregard for the institute in seeking to take advantage of my awards. One or two of them felt it would be better if I simply resigned and left the institute entirely. Dr. Mekako felt sympathetic and in a very private discussion supported whatever decision I made. He was prepared to assist me in whatever way possible to retain my job especially as "the institute needed your expertise and newly acquired experience." In his opinion, RRIN will be losing one of their most accomplished young research scientists. Personally, I have always believed that no one was indispensible considering we had opened a number of the first group of rubber clonal trials for tapping against the suggestions of some of our senior research scientists when they were out of the country. The clones selected eventually became the first group of NIG series of clones for RRIN. After much discussion coupled with a few more visits from Mr. Agha and a lot of agonizing, my wife and I, decided in the interest of the future of our nuclear family we would take advantage of the awards and proceed overseas to Iowa State University for my Ph.D. We prayed that God Almighty would take care of everyone including my family who was actually making a critical decision that will impact us for the rest of our lives. Thus by the end of June 1981 a final decision was made and we started full preparation for me to travel to the USA in early August. With the backdrop of "Germplasm 81" still fresh in the minds of my superiors, I applied for study leave without pay. I expected a negative response. Meanwhile, I continued to work as hard on the many research projects even as I intensified my preparation for leaving the country.

In the midst of all that was going on I was pleased when one morning I went to the germplasm nursery, and saw Mr. Aransomwan and Mr. Anetor, my agricultural superintendent (AS) and assistant (AA) respectively smiling and waving frantically to me as they showed me all the new seedlings which have resulted from the seeds I brought from Brazil. I could not believe the success rate of germination for the seeds. As trusted technicians, I broke the news to them of my impending departure for the USA to study for my Ph.D. To my amazement they took the news very well. In fact they congratulated me and said they knew I was going to be moving on giving the way Dr. Mekako always talked to them about me. "We have been working with you for nearly five years and we always knew that your future was very bright. See how much you have done for the institute in a very short time. We pray to God to bless you and your family." Both of them concluded. I thanked them immensely for all the support they had given me from the moment I arrived as a research officer in training (ROT) at RRIN especially how they helped me work with daily labor to establish new clonal trials in three natural rubber producing states. I wished them all the very best and concluded by saying "I am leaving these plants in very good hands. I am confident this will be an important nursery for the institute." Shortly after this nursery visit to see one of the outcomes of my being in the

Amazon Basin I sold my new car that had arrived six weeks earlier and was operational. I purchased a ticket to fly on the Nigerian Airways flight directly to New York and on to Ames, Iowa in the first week of August 1981.

# EPILOGUE

The reunion in Manaus revealed that "Germplasm 81" was very successful in accomplishing the goals set out for the team of IRRDB and Brazilian scientists. As a retrospective, the expedition was quite extensive and expensive. It is questionable whether such a huge expenditure could be conducted again given the political, social and financial conditions that now exist in the Amazon basin. Yet, *Hevea brasiliensis* is presently the only commercial source of natural rubber, making it a very important strategic commodity not only for the United States, but also for countries around the world. *Hevea's* major economic importance lies in its sap-like extract, also known as latex, the primary source of natural rubber. Rubber has a widespread use, spanning from the manufacture of household items over medical to industrial goods. Natural rubber has been used as an intermediate or final component in the production stream of most rubber products, such as gloves, balloons, adhesives, hoses, etc. Tires and tubes make up rubber's largest market share. Other significant uses of rubber include the paper, carpet and textile industries, where rubber's low tenacity and oxidizing resistance limits its use in, for example, lightweight garments, substituting it thereby with neoprene or synthetic rubber. Natural rubber, however, due to its unusually high molecular weight, is far superior to synthetic rubber, and cannot be substituted for the manufacturing of many products, such as aircraft, space shuttle and heavy construction equipment tires like those of Caterpillar and Halliburton, which require 100% natural rubber.

Today, ninety percent of the world's natural rubber comes from rubber tree plantations in a few countries (Malaysia, Indonesia, India, Thailand and Vietnam) in Southeast Asia (West Africa has nine percent, while Brazil and other Latin

American countries produce one percent). The vulnerability of the rubber market has prompted Congress to mandate the development of new commercial sources of natural rubber. However, to date, no new replacement sources or adequate substitution have been found, making natural rubber supply an international security issue, with the potential for destabilizing the world economy. For example, a natural rubber shortage could easily result in a complete halt in the manufacturing of airplane tires. And with the aviation industry being a key driver of the world economy, such situation could be proven disastrous. In reality, this scenario is not far-fetched, considering the damaging potential of biological agents such as the South American Leaf Blight. This fungal disease caused by the ascomycete *Microcyclus ulei*, single handedly managed to prevent natural rubber production on a commercial scale in South and Central America, despite the efforts of Goodyear and Firestone multinational companies in the late 1800s and early 1900s. Although today, the spread of this pest is still limited to South America, the potential of an economic disaster increases with every transcontinental flight landing in Southeast Asia.

Imagine for a moment, involving the release of *Microcyclus* spores in a *Hevea* plantation in Thailand or Indonesia or Malaysia where 80% of the world rubber is produced! The pathogen can be easily isolated from infected rubber trees in Brazil and transported undetected across borders. Transfer can also occur through host (budwood, foliage, flowers, fruit and seeds, in-vitro plants) and non-host materials (contaminated goods). Depending on the material, the viability of conidia or spores could be compromised during transport. However, considering the similarities in climatic and environmental conditions between Southeast Asia and the pathogen endemic region in Brazil, the likelihood of an invasion increases tremendously. There are also numerous published reports on non-technical fungus isolation and storage, where terrorists can take notes. The potential misuse of print cartridges to diffuse lethal airborne substances to a massive population of computer users around the world, where common substances are mutated into very lethal weapons has been reported in the past. It has also been demonstrated that fungal spores such as those responsible for the South American leaf blight (*Microcyclus ulei*) when stored dry in glass vials have the capacity to survive and germinate up to one month, while they will propagate after a year when stored on wet agar plates.

Although decades of breeding have resulted in highly selective rubber tree cultivars or clones, nearly all rubber tree plantations around the world are descendants of seeds originated from the 1876 Sir Henry Wickham Collections from a single location in Para State (hence para rubber) of Brazil, and that survived in the shipment from Kew Gardens, England to Asia. This narrow genetic background makes these rubber tree plantations particularly vulnerable to diseases and insect pests as well as unpredictable environmental conditions. Over the years, rubber tree improvement has resulted in varieties and clones with traits that give the highest return, such as the quantity and quality of latex harvested. However, selection processes have disturbed natural rubber balance and created a so-called genetic homogeneity (inbreeding). Crop genetic uniformities are usually welcomed, as most of them expand superior resistance to pest and diseases. Nonetheless, over the centuries, pests have developed the capacity to overcome host resistance by genetically evolving and gaining access to larger uniformed crop base plantations where damages could be severe. The concerns of genetic uniformity became apparent in 1970, when the spread of the Southern Corn Leaf Blight (*Helminthosporium maydis*) led to a 15% reduction in the U.S. corn crop. This issue of genetic vulnerability is a broadly recognized problem affecting genetic resources globally. Consequently, significant efforts to collect and preserve genetic resources have been made by national and international institutions as an insurance policy against future disasters. The recognition, however, that *Hevea brasiliensis* plantations around the world originally stem from a single source raises new concerns that are global in scope.

Travel methods (sea or air) usually determine the entry potential of a pest. Although fungal spores will have a high level of survival if air-transported through fresh host materials such as leaves, the duration and environmental conditions of sea cargo would not promote survival and viability of the pathogen. At the port of entry, the probability of not being properly identified is enormously high, as pathogen detection through visual inspection alone is almost impossible. For example, If properly concealed, budwood has the potential to act as both a vector and a host. Once established, the pathogen can cause severe economic damages, as seen in Brazil, where plantations were completely abandoned in the past making Brazil to import over 60 percent of its natural rubber currently.

The very narrow genetic base of the commercial natural rubber clone plantations in Southeast Asia makes them extremely susceptible to Leaf Blight infection. Within a few years following invasion, all standing stock in the region will most likely be infected and destroyed, resulting in unforeseen direct and indirect economic and environmental consequences worldwide. A wide spread infestation would require costly control measures, which in the long run would not be economically sustainable. An uncontrolled infestation on the other hand would result in trees dieback and concomitant loss of vast plantations, which will directly affect the livelihood of hundreds of smallholders and indirectly decimate the natural rubber industry. Economically, this could be translated as a loss of revenue in the amount of over $100 billion annually, significantly impacting both producing and importing countries alike and crippling the overall world economy!

Fortunately, in the history of humankind, there are no known reports of a global intentional biological invasion. The deliberate use of anthrax in 2001 was limited to the United

States alone. Subsequently however, this incident resulted in a substantial increase in funding to counter the threat of bioterrorism the world over. Global initiatives to detect and respond to biological threats were suddenly becoming increasingly vital to most nations. Worldwide models and scenarios were constructed and consequently attacks potentially disruptive to global supply chains and human welfare, remain exceedingly low. More importantly, new advances in science and technology have helped us gain ground against these threats. The challenges presented by *Hevea brasiliensis* however remain. Today, there is an immediate need to speed up the development and production of alternatives to *Hevea brasiliensis* through translational research. Although gayule (*Parthenium argentatum*) and gutta-percha (*Palachium gutta*) have been attracting attention as sources of hypoallergenic rubber for more than 50 years, their limited geographic range appears to restrict their expansion. More research investments for example, could target these two crops as possible alternatives to *Hevea*, as many sectors recognize that when it comes to upstream planning and prevention efforts against bioterrorism, the benefits outweigh the costs. On the other hand it may be necessary to strengthen and enhance research efforts on *Hevea* in existing rubber producing countries or provide innovative approaches which could restore the natural rubber industry to Brazil as a necessary buffer to any potential crisis in the loss of natural rubber supply should the Asian plantations be lost to *M. ulei.*

It is over 30 years since "Germplasm 81" was completed. This unique expedition was very intensive and expensive. No other expedition of the same size has been undertaken since then. From the stand point of new collections in order to further broaden the genetic base of rubber clones in commercial plantations, collections in Mato Grosso will have to visit some locations which the 1981 team could not access. These include, the primary forest area of Rio Juruena which revealed that almost each collocacao had its own genetic resource of *Hevea.* Second area to be used is the Rio Tartaruga region that was said to contain more than 1,500 seringueros. The area is about 12 miles from Aracatuba. Because it represents a transitional zone between the rainforest and savanna grassland (cerrado), it will be a very useful source of materials for genetic presevation. Another area for *ex situ* germplasm collection in Mato Grosso is the Utiariti location considered to be native Indian territory in Brazil. Collection in this region will complete all the geographical areas of Mato Grosso. As indicated earlier, it is questionable whether a huge expenditure as was done for "Germplasm 81" could be conducted again given the political, social and economic conditions that now exist in the Amazon basin. Therefore*, in situ* germplasm preservation may be the route to take in the future. This is now an ongoing activity of the United States National Plant Germplasm System. Growing *Hevea* in its native habitat ensures longevity and compliments efforts in ex situ locations such as gene banks. Some areas in Brazil called "Amazon Natural Rubber Conservation Areas (ANRCs) could be set aside and later used as germplasm sources. Perhaps none monoculture rubber plantations could be established using sustainable tree crop agroforestry systems in which improved rubber clones are grown in between rows of Brazilian nut trees and other important economic tree/fruit crop species to mimic conditions in the rainforests. This will ensure that *M. ulei* spores do not readily spread from tree to tree as the food source is cut off. Finally, the use of new genetic engineering tools to develop biological control agent, for the management of this devastating natural rubber disease will be highly beneficial. Bacteria or viruses or even other fungal strains could be genetically engineered to biologically control the spread of the blight not only in Brazil but also in all natural rubber producing countries in Southeast Asia and West Africa. It may also be interesting to consider evaluating some rubber clones in Key

West and Homestead areas in Florida for possibilities of growing what one may call "Distance Buffer Natural Rubber Plantations (DBNRP)." Natural rubber is such an important commodity whose shortage could make the shortage of crude oil look like "child's play" when compared to the disruption that will occur throughout the world's economy. Given the gestation period of the rubber tree to reach tapping age and start producing the latex (white gold) which is eventually processed to so many valuable products, it is critical that every effort be made to ensure that genetic materials be constantly grown and/or preserved to avoid any doom's day scenario.

"Germplasm 81" must be considered a major accomplishment for the International Rubber Research and Development Board in which member institutes were able to pool resources together to send their scientists to undertake such a unique expedition in the center of origin of a plant species. The Brazilian government, Brazilian scientists and other Brazilians in the far reaches of the Amazon Basin who welcomed foreign scientists at every point must be highly commended. As an African Scientist in the Amazon Basin this was a lifetime experience in which I was able to see the remotest parts of the tropical rain forest of Brazil especially in the State of Mato Grosso. Seeing a rubber seed on the forest floor of its native habitat in the Amazon Basin has perhaps been the most rewarding experience of my professional career. The human relationships in many cases are onetime experiences as well but remain memorable. The challenge is always to keep such relationships over a long period of time. But these are memories that remain with us throughout our lives. This has been the case for me. Furthermore the experience gained has been instrumental to the success of my professional career in international research and development for more than thirty years. A sustainable natural rubber industry is *sine qua non* to the success of the global economy and the well being of all humans.

# ABOUT THE AUTHOR

Dr. Oghenekome U. Onokpise (pen name: Kome Onokpise), is a Full Professor and Associate Dean in the College of Agriculture and Food Sciences, Florida A&M University, Tallahassee, Florida, where he has been teaching, conducting research and outreach activities especially international development for more than twenty five years. He has also worked at the Rubber Research Institute of Nigeria, Iyanomo, Benin City, Edo State, Nigeria, and served as a pioneering National Youth Service Corps member at the School of Agriculture, Yandev, Gboko, Benue State, Nigeria. He also holds courtesy professorship in the School of Forest Resources Conservation, University of Florida, Gainesville, Florida. He has received many teaching and research awards including a patent from the U.S. Patenting Office.

Born in Lagos, Professor Onokpise was raised by his grandparents until he was eight years old at a small village, Otha (small Ughelli), near Lake Ebe, Ovrode in Isoko North local government, Delta State, Nigeria. He started his elementary school at Lake Ebe/Igbuku elementary schools. He joined his parents at the age of eight years old and completed his elementary school education at St. Mary's Catholic School, Lokoja, Kogi State, Nigeria. He attended and completed his secondary school education at the elite Government College Ughelli, Delta State, Nigeria, where he obtained his Western African School Certificate (WASC). Thereafter, he attended the Federal of School Science, Onikan, Lagos State, Nigeria and obtained the General Certificate of Education, Advanced ("A") Level (GCE, A, level), administered by the University of London in the United Kingdom. He obtained his B.Sc. in General Agriculture from the University of Ife (Obafemi Awolowo University), Ile-Ife, Oyo State, Nigeria; his M.Sc. in Plant Breeding and Genetics from the University of Guelph, Guelph,

Ontario, Canada; and his Ph.D. in Tree Breeding and Forest Genetics from Iowa State University, Ames, Iowa, USA; and did his post-doctoral research work at Ohio Agricultural Research and Development Center (OARDC), The Ohio State University, Wooster Campus. Ohio, USA.

Kome Onokpise is married to his wife for 35 years. They have been blessed with four sons and one grandson. Kome Onokpise is a naturalized citizen of the United States of America (a Diosporan Nigerian). A devout Catholic, Kome is a former altar boy of St. Mary's Catholic Church, Lokoja, Kogi State; and St. Peter and Paul's Catholic Church, Ughelli, Delta State. He is presently a member of the Choir and Knights of Columbus at St. Louis Catholic Parish, Tallahassee, Florida. Professor Onokpise is a member of a number of professional organizations such as the Society of American Foresters (SAF), American Association for the Advancement of Science (AAAS), Society for Economic Botany (SEB), etc. Kome is also a member of Tallahassee Southside Rotary Club International and a member of the Tallahassee Writers Association (TWA). He enjoys reading, singing, traveling, walking and writing. He has been to five continents including very interesting trips to the Amazon Basin, Brazil; Rumpi Mountains, Cameroon; Gottingen, Germany; and Kuala Lumpur, Malaysia. He has published many scientific journal papers and book chapters. He is the author of *"Growing Up: Tony Joins the System."* He is also co-author of *"Soil Science for Archeologists"*, and *"Niger Delta Recipes: Preserving Traditional Spices and Condiments for Healthy Living."*

# UPCOMING BOOKS

1. **PAN AM in Africa: My First Flight – Lagos to Kuala Lumpur.** *In 1977, Kome Onokpise was selected to attend an international conference in Kuala Lumpur, Malaysia. This was going to be his first overseas travel. There were other airlines like the British Airways, KLM, Swissair and Lufthansa flying the West African region. For some reason unknown to both the Director of RRIN and Kome Onokpise himself, his flight was booked on PAN AM the only American airline flying through the region at the time. Kome loses his father in a fatal car accident which occurs only eight days before undertaking his first overseas flight on PAN AM. The decision to travel, his experience on PAN AM and his three weeks stay in Kuala Lumpur as a 26 year old who just lost his father, and attending a major conference on behalf of his institute and country in the midst of very experienced and accomplished scientists, are covered in this new book.*

2. **The Rumpi Mountains: The Story of a USAID Project in Cameroon.** *From 1987 to 1994, the United States Agency for International Development (USAID) and the government of the Republic of Cameroon funded a research project in the amount of eight million dollarS ($8.00 m) for the development of new cocoyam varieties for the enhancement of food security and the general well being of the people of Cameroon. This was*

*the first and largest funded project to a consortium of three 1890 Land Grant Universities (University of Maryland Eastern Shores – UMES, Alabama A&M University - AAMU and Florida A&M University – FAMU) in the United States. Kome Onokpise represented Florida A&M University on a two-year long-term assignment. He was made the team leader for the Cocoyam Breeding Team. With his experience from "germplasm 81" he was asked to lead a team to collect local cocoyam germplasm in four provinces of Cameroon. His team consisted of two female technicians, one of whom was six months pregnant while the second female was a single parent. Their expedition took them up into the RUMPI Mountains of Cameroon. At the top of the mountain which is 800 meters (1,300 feet) above sea level, they found a cocoyam farm that had been farmed continuously for almost 400 years. This experience and what happened during Kome's long-term tour of duty is covered in this thriller.*

3. **Crazy Curly Corn: The Mutants.** *This is a book for high school students that pictorially discusses how corn hybrids and varieties are developed. It takes the difficult topics of mutation and corn mutants and simplifies them for ninth to twelfth graders thus enhancing their knowledge of genetics. The book is based on experience gained over two funding cycles by a National Science Foundation Grant to a group of Principal Investigators at Florida State University and Florida A&M University respectively. The book represents the outreach portion of the grant to minority students as a way to encourage them to get*

*interested in the science of genetics. With pictures of students in corn fields cross pollinating plants and other demonstrations, the book brings home an easy to read document for students in high schools and the general public.*

Kome was, in his youth, and continues to be an avid reader.

Kome Onokpise, his wife Lucy, and their first son.

Spots pm leaves caused by the South American leaf blight disease.

Trekking through the Amazon Jungle.

Several men standing near a native rubber tree, where latex cups can be seen.

Mato Grosso, Brazil, a Germplasm collection area.

Germplasm 81 Collection Team from member Institute of IRRDB and their Brazilian counterparts.

Dr. Onokpise's Rubber Seed Journey Germplasm 81
January 25, 26, 1981 – April 28, 1981

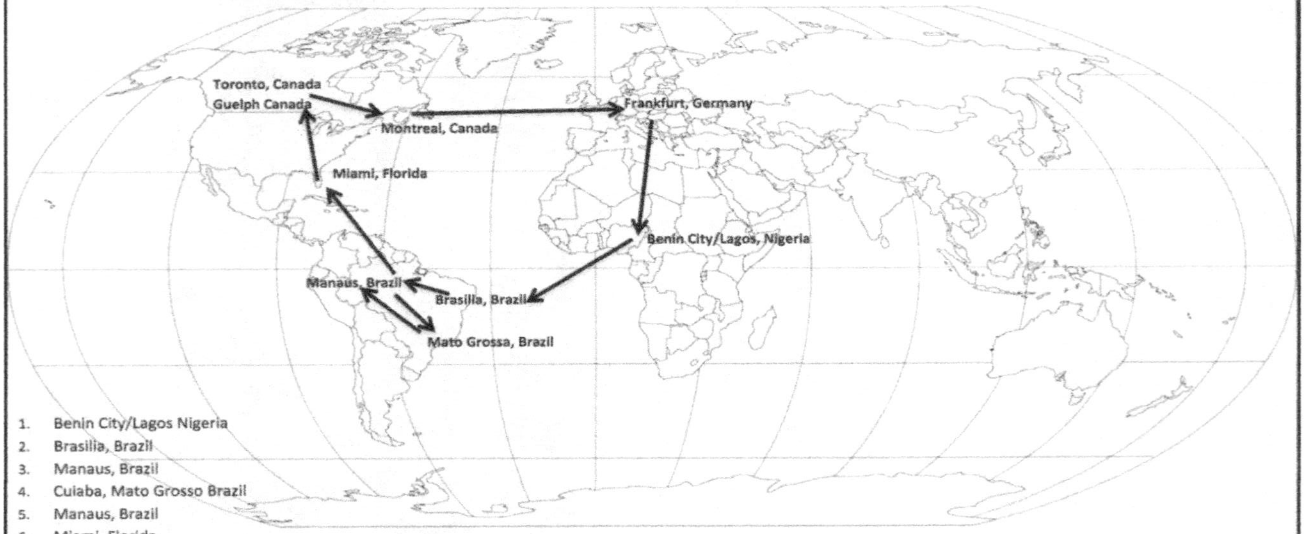


1. Benin City/Lagos Nigeria
2. Brasilia, Brazil
3. Manaus, Brazil
4. Cuiaba, Mato Grosso Brazil
5. Manaus, Brazil
6. Miami, Florida
7. Toronto, Canada
8. Guelph, Canada
9. Montreal, Canada
10. Frankfurt, Germany
11. Lagos, Nigeria
12. Benin, Nigeria

Kome Onokpise traveled from Nigeria and back, landing on four continents during a four month period. The above map follows his journey, departing Lagos, Nigeria on January 31, 1981 and back home on April 28, 1981.

Made in United States
Orlando, FL
11 November 2026

90609406R00105